CHARACTERIZATION OF EXHAUST EMISSIONS FROM HEAVY-DUTY DIESEL VEHICLES IN THE HGB AREA – FINAL REPORT

by

Jeremy Johnson
Associate Research Specialist
Texas Transportation Institute

Doh-Won Lee
Assistant Research Scientist
Texas Transportation Institute

Reza Farzaneh, Ph.D., P.E.
Assistant Research Engineer
Texas Transportation Institute

Josias Zietsman, Ph.D., P.E.
Division Head
Texas Transportation Institute

and

Lei Yu, Ph.D., P.E.
Dean, College of Science and Technology
Texas Southern University

Report 0-6237-1
Project 0-6237
Project Title: Characterization of Exhaust Emissions from Heavy-Duty Diesel Vehicles in the HGB Area

Performed in cooperation with the
Texas Department of Transportation
and the
Federal Highway Administration

October 2011
Published: February 2012

TEXAS TRANSPORTATION INSTITUTE
The Texas A&M University System
College Station, Texas 77843-3135

DISCLAIMER

This research was performed in cooperation with the Texas Department of Transportation (TxDOT) and the Federal Highway Administration (FHWA). The contents of this report reflect the views of the authors, who are responsible for the facts and the accuracy of the data presented herein. The contents do not necessarily reflect the official view or policies of the FHWA or TxDOT. This report does not constitute a standard, specification, or regulation. The engineer in charge of this project was Josias Zietsman, Ph.D., P.E., (Registration - TX #90506).

ACKNOWLEDGMENTS

This project was conducted in cooperation with TxDOT and FHWA. The authors wish to thank the Project Director, Tim Wood, and PMC members including Duncan Stewart, Jackie Ploch, Ruben Casso, Charles Airiohuodion, Don Lewis, Graciela Lubertino, Jim Price, Morris Brown, Madhu Venugopoal, Paul Tiley, and Shelley Whitworth for their input and guidance. The authors also wish to thank the City of Houston for their support and for providing access to their fleet vehicles during the testing process.

TABLE OF CONTENTS

Page

List of Figures ix
List of Tables xi
List of Acronyms xii
Executive Summary 1
Chapter 1: Background and Introduction 3
   Project Need 3
   Overall Goal 3
   HGB Air Quality Status and Mitigation Strategies 4
      Air Quality Status 4
      Mitigation Strategies 5
      Mobile Source Air Toxics 5
   MOVES 6
Chapter 2: Project Approach and Test Protocol 9
   Overall Approach 9
   Drive Cycle Development 12
      Drive Cycles for Emissions Testing 12
      Drive Cycles for Emissions Analysis 15
   Test Facilities and Equipment 16
      TTI Environmental and Emissions Research Facility 16
      Test Equipment 17
      SEMTECH-DS 17
      Axion 18
      Microdilution Sampling System 19
      Dekati Mass Monitor 20
Chapter 3: Test Vehicle Selection 21
   Target Vehicles 21
      HEs 21
   Vehicle Selection Process 22
      Fleet Selection 22
      HEs Selection 22
Chapter 4: Testing Results and Analyses 25
   Driving Testing Results 26
      Modal Emissions 27
   Impact Analysis and Comparison with MOVES 31
      Overall Results 37
   Idling Testing Results 38
      Class 8 HEs 38
      Class 8 Randomly Selected Vehicles 41
      Class 6 Vehicles 43
      Class 4 HEs 46
      Overall Results 48
   Emissions Reduction Benefit Analysis 50
   Correlation of Different PM Measurements 51

Chapter 5: Conclusions 55
References 59
Appendix A: Driving Emissions Result Comparisons with MOVES Estimates 61
**Appendix B: MOVES-Based Algorithm for Estimating the Benefits of Upgrading a
Potential High Emitter Vehicle** 63

LIST OF FIGURES

Page

Figure 1. Total VMT and NOx Emissions for Selected Vehicle Classes in Harris County. 4
Figure 2. Relationship between MSAT and Diesel-Related Compounds. 6
Figure 3. Task Flow Chart. 10
Figure 4. Class 4 Vehicle Instrumented for Testing. 11
Figure 5. Sample Daily Speed Profile for a HDDV 8 Vehicle. 15
Figure 6. Using Real-World GPS Data for Emissions Analysis. 16
Figure 7. City of Houston Vehicle inside EERF Chamber. 17
Figure 8. SEMTECH EFM Installed on Test Vehicle (Left) and SEMTECH-DS (Right). 18
Figure 9. CATI Axion System along with the SEMTECH-DS Installed on Test Vehicle prior to Testing. 19
Figure 10. Microdilution Sampling System. 20
Figure 11. Dekati Mass Monitor. 20
Figure 12. Pictures of Opacity Testing on COH Fleet Vehicles. 23
Figure 13. Cumulative Frequencies of Opacity Readings. 24
Figure 14. Average Modal Emission Rates for Vehicle 8H-2 (21832). 28
Figure 15. Average Modal Emission Rates for Vehicle 6-2 (29332). 29
Figure 16. Average Modal Emission Rates for Vehicle 4H-4 (29417). 30
Figure 17. Example Drive Cycles: Test Cycle (Left), Operation Cycle (Right). 32
Figure 18. Total Emission for the Test Cycle of Vehicle 8H-2 (21832). 33
Figure 19. Total Emission for Operation Cycles of 8H-2 (21832). 34
Figure 20. Cycle Analysis Results for High Emitting Class 8 Vehicles. 36
Figure 21. Cycle Analysis Results for Randomly Selected Class 8 Vehicles. 36
Figure 22. Cycle Analysis Results for Class 6 Vehicles. 37
Figure 23. Cycle Analysis Results for High Emitting Class 4 Vehicles. 37
Figure 24. Class 8 HEs Idle Emissions. 38
Figure 25. Class 8 HEs PM Emissions. 39
Figure 26. Class 8 HEs Aldehyde Emissions. 39
Figure 27. Class 8 Randomly Selected Vehicles Idle Emissions. 41
Figure 28. Class 8 Randomly Selected PM Idle Emissions. 42
Figure 29. Class 8 Randomly Selected Aldehyde Idle Emissions. 42
Figure 30. Class 6 Idle Emissions. 44
Figure 31. Class 6 PM Idle Emissions. 45
Figure 32. Class 6 Aldehyde Idle Emissions. 45
Figure 33. Class 4 HEs Idle Emissions. 46
Figure 34. Class 4 HEs PM Idle Emissions. 46
Figure 35. Class 4 HEs Aldehyde Idle Emissions. 47
Figure 36. Comparison of Emissions Rates of Vehicle Classes. 48
Figure 37. Comparison of Aldehyde Emissions Rates of Vehicle Classes. 49
Figure 38. Comparison of PM Emissions Rates of Vehicle Classes. 49
Figure 39. Correlation between PM Filter Emission Rates and Opacity Readings. 52
Figure 40. Correlation between Filter and DMM PM Emission Rates (Class 8H Vehicles). 53
Figure 41. Correlation between Filter and DMM PM Rates (Class 8 Vehicles). 53

Figure 42. Correlation between Filter and PEMS PM Rates (Class 8H Vehicles). 54
Figure 43. Correlation between Filter and PEMS PM Rates (Class 8 Vehicles). 54

LIST OF TABLES

Page

Table 1. MOVES Operating Mode Bin Definitions for Running Emissions. 13
Table 2. Summary of Vehicles in GPS Data Collection. 15
Table 3. HGB Area – Select HDDV Classes. 21
Table 4. Selected Vehicles Opacity Results. 24
Table 5. Tested Class 8 HEs. 25
Table 6. Tested Class 8 Randomly Selected Vehicles. 26
Table 7. Tested Class 6 Vehicles. 26
Table 8. Tested Class 4 Vehicles. 26
Table 9. MOVES' Vehicle Parameters Values. 27
Table 10. Characteristics of Operating Cycles – Class 8 Vehicles. 32
Table 11. Characteristics of Operating Cycles – Class 6 Vehicles. 32
Table 12. Characteristics of Operating Cycles – Class 4 Vehicles. 32
Table 13. Observed Operation Cycles Emission Rates; High Emitting Class 8 Vehicles. 34
Table 14. Observed Operation Cycles Emission Rates; Randomly Selected Class 8 Vehicles. .. 35
Table 15. Observed Operation Cycles Emission Rates; Class 6 Vehicles. 35
Table 16. Observed Operation Cycles Emission Factors; High Emitting Class 4 Vehicles. 35
Table 17. Class 8 HEs Percentage of MOVES Emissions. 40
Table 18. Class 8 HEs Percentage of MOVES Emissions. 43
Table 19. Class 6 Percentage of MOVES Emissions. 44
Table 20. Class 4 HEs Percentage of MOVES Emissions. 47
Table 21. MOVES Rates Used for Emissions Reduction Analysis - Analysis Year 2011. 50
Table 22. Potential Emissions Benefits by Replacing MY 2000 or Earlier Vehicles. 51
Table 23. Potential Fuel Savings and Emissions Benefits by Replacing HEs with Normal Vehicles. 51

LIST OF ACRONYMS

1. Average Annual Miles – AAM
2. Baseline Emissions Rates – BER
3. California Air Resources Board – CARB
4. Carbon Dioxide – CO_2
5. Carbon Monoxide – CO
6. City of Houston – COH
7. Congestion Mitigation Air Quality – CMAQ
8. Dekati Mass Monitor – DMM
9. Electrical Low Pressure Impactor – ELPI
10. Emissions Adjustment Factor – EAF
11. Emissions Reduction Incentive Grants – ERIG
12. Environmental and Emissions Research Facility – EERF
13. Environmental Protection Agency – EPA
14. Exhaust Flow Meter – EFM
15. Federal Highway Administration – FHWA
16. Flame Ionization Detector – FID
17. Global Positioning System – GPS
18. Gross Vehicle Weight Rating – GVWR
19. Heavy Duty Diesel Vehicle – HDDV
20. High Emitting Vehicle – HE
21. Houston Exposure to Air Toxics Study – HEATS
22. Houston-Galveston Area Council – HGAC
23. Houston-Galveston-Brazoria – HGB
24. Hydrocarbons – HC
25. Inspection/Maintenance – I/M
26. Light Duty Gas Vehicles – LDGV
27. Methane – CH_4
28. Microdilution Sampling System – MSS
29. Mobile Source Air Toxics – MSAT
30. Model Year – MY
31. Motor Vehicle Emissions Simulator – MOVES
32. Nitrous Oxide – N_2O
33. Nonattainment – NA
34. Oxides of Nitrogen – NOx
35. Particulate Matter – PM
36. Portable Emissions Measurement System – PEMS
37. Relative Humidity – RH
38. Selective Catalytic Reduction – SCR
39. Society of Automotive Engineers – SAE
40. Texas A&M University – TAMU
41. Texas Commission on Environmental Quality – TCEQ
42. Texas Department of Transportation – TxDOT
43. Texas Emissions Reduction Plan – TERP
44. Texas Southern University – TSU

45. Total Emissions Reduced – TER
46. Total Gaseous Hydrocarbons – THC
47. Vehicle Miles of Travel – VMT
48. Vehicle Specific Power – VSP
49. Volatile Organic Compounds – VOC
50. Voluntary Mobile Emissions Reduction Programs – VMEP

EXECUTIVE SUMMARY

Many areas in Texas, including the Houston-Galveston-Brazoria (HGB) area, face air quality issues and are in nonattainment of federal ambient air quality standards. In these areas, state and local transportation agencies seek to implement various strategies to reduce mobile-source (i.e., vehicular) emissions (*1*).

During the previous decade the contribution of heavy-duty diesel vehicles (HDDVs) to overall mobile source emissions has greatly increased. This increase is due to various factors, including growing freight volumes resulting in increased HDDV movement and reduced emissions from light-duty vehicles (LDVs) due to their improved engine technology. This increase in emissions from HDDVs makes them an important target for programs aimed at reducing emissions. Previous research has also indicated that certain vehicles identified as "high emitters" (HEs) contribute disproportionately to the overall HDDV emissions. Identifying HEs and characterizing their emissions is another aspect that is important for targeted emissions reduction initiatives.

In this project, the research team studied the emissions characteristics of different classes of HDDVs operating in the HGB area, with a view of understanding and characterizing their emissions. This project focused on three classes of HDDVs (referred to as Classes 4, 6, and 8b); each class has different vehicle weight ratings. The emissions characteristics of potential HEs were compared against vehicles representative of the general (normal emitting) HDDV fleet. Additionally, the measured emissions were compared to rates estimated using the MOtor Vehicle Emission Simulator (MOVES) emissions model to study differences and trends.

Prior to performing emissions testing, the research team searched potential vehicle fleets for the emissions testing to identify a fleet that is representative of the overall population of HDDVs in the HGB area. The vehicle fleet belonging to the City of Houston (COH) was identified as the best candidate and selected to participate in the testing. Among the selected classes of COH HDDVs, the team identified HEs by conducting opacity testing. Over 90 vehicles in the COH fleet were screened using opacity testing, and a set of 12 HEs, six from Class 4 and six from Class 8, were selected for the emission testing.

Using a total of 30HDDVs, which included the 12 HEs, the research team performed both driving and idling emissions testing. The driving testing was performed by following drive cycles which were also developed for those vehicles during this study. Due to the nature of various driving characteristics such as different vehicle speeds, accelerations, and engine loads, driving testing results were mainly used for comparisons of MOVES estimates. The idling testing was performed in a simple test condition, which is designed mainly for comparisons among vehicle classes/types.

In general, it was observed that vehicle emissions vary greatly even within a vehicle class, and vehicles identified as HEs differed in their emissions characteristics from the randomly-selected vehicles. For example, the Class 8 vehicles identified as HEs showed differences when compared to the randomly-selected vehicles; in the case of idling, the HEs consumed 21 percent more fuel than the randomly selected test vehicles and produced between 24 percent and

87 percent higher levels of pollutants (depending on the pollutant types). Similar results were observed for the driving emissions as well. Due to limitations of sample size and differences in test conditions, the usefulness of opacity testing as a means to identify HEs could not be statistically tested.

The results from this project demonstrate that a viable emissions reduction strategy could be to screen HE vehicles from the fleet and replace them or install emissions control technologies for maximizing the emissions reduction and air quality benefits. Larger vehicle fleets, especially those with older vehicles in the HGB area and other nonattainment (NA) areas, can provide many opportunities to apply these strategies for regional air quality improvement.

CHAPTER 1: BACKGROUND AND INTRODUCTION

PROJECT NEED

Because the HGB area is an eight-hour ozone NA area, state and local transportation agencies in the HGB area seek ways that eliminate and/or reduce emissions from various sources including mobile sources. In order to eliminate and/or reduce emissions effectively, it is important to know emissions contribution from each component of the source. The relative contribution of HDDVs to mobile source emissions has grown significantly over the past decade. It is critical to address this component of the fleet in the HGB eight-county ozone NA area to mitigate emissions effectively from the component.

Most emissions studies have not incorporated random sampling in their study designs. Also, they are mostly based on laboratory settings using chassis dynamometer testing and are focused on gaseous pollutants such as hydrocarbons (HC), carbon monoxide (CO), and oxides of nitrogen (NOx). That is, most of the studies do not include particulate matter (PM) and mobile source air toxics (MSAT) into their studies. To the best of their knowledge, the research team found no study that addresses PM and MSAT as well as gaseous emissions and incorporates real world testing and random sampling into the study. These components are very important. Without random sampling, the results can always be biased. Both PM and MSAT have been identified by the U.S. Environmental Protection Agency (EPA) having a critical effect on health and must be investigated. In addition, the latest federal regulations require in-use measurement of emissions (2). This project addresses all of these aspects as well as the component that is often overlooked—the emissions impact of HEs.

OVERALL GOAL

The main goals of this project were:
- To characterize emissions from different classes of HDDVs operating in the HGB area.
- To identify HEs and characterize emissions from HEs.
- To compare emissions of HEs and non-HEs.
- To compare test results with estimates from MOVES (EPA's emissions model).

The research team recruited a HDDV fleet in the HGB area, which operated in the HGB area, and performed emissions testing on the vehicles, including targeted HEs. The test results were processed, compared, and analyzed to determine the overall impact of different classes of HDDVs and HEs, how to identify the HEs, how the obtained results compare to the estimated emissions rates from the MOVES model, and the potential benefits if HEs are replaced with normal emitting vehicles in the HGB area.

HGB AIR QUALITY STATUS AND MITIGATION STRATEGIES

Air Quality Status

For this study, the HGB area was selected because its status is designated as severe nonattainment for ground-level ozone under the eight-hour ozone standard (*3*). The information gathered on HDDV emissions in this area can be applicable to other areas in Texas. Counties affected in the HGB area are Brazoria, Chambers, Fort Bend, Galveston, Harris, Liberty, Montgomery, and Waller. Meeting the ozone standard is especially challenging for the HGB region due to its unique meteorological conditions, complex ozone formation chemistry, and the magnitude of reductions required. Control strategies mentioned in the State Implementation Plan (SIP) include Federal On-Road Measures, Vehicle Inspection/ Maintenance (I/M), Speed Limit Reduction, Cleaner Diesel, Voluntary Mobile Emission Reduction Programs (VMEP), and Transportation Control Measures (*4*). Some of the measures under VMEP are vehicle scrappage and smoking vehicle programs. The results contained in this report make it possible to quantify the benefits of actions taken under these programs.

Figure 1 is based on the Texas Transportation Institute's (TTI's) conformity determination work for Harris County. It illustrates that even though HDDVs (especially Class 8b HDDVs) have far fewer vehicle miles of travel (VMT) than light duty gas vehicles (LDGVs), their contribution to NOx emissions are by far the greatest. In the case of mobile source PM emissions, by far the majority is caused by HDDVs (*5*).

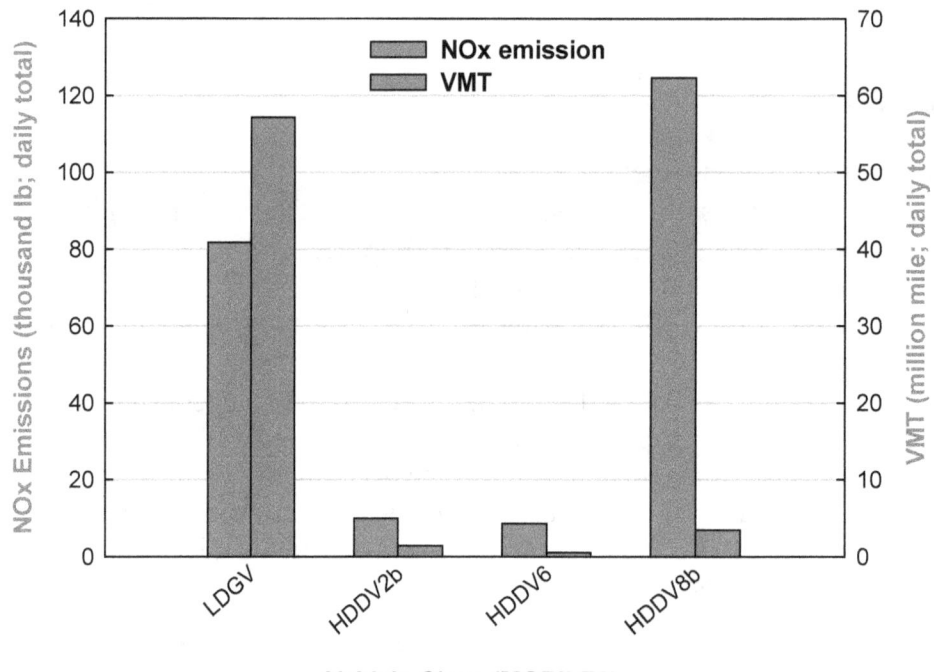

Figure 1. Total VMT and NOx Emissions for Selected Vehicle Classes in Harris County.

Mitigation Strategies

Texas has a financial assistance and incentive program for qualified owners of gasoline-powered vehicles that fail the emissions test or are at least 10 years old. However, diesel-powered vehicles are exempt (*4*). The vehicle scrappage (repair and replacement) program is officially called *AirCheckTexas Drive a Clean Machine* and applies only in participating counties in areas of Texas that have vehicle I/M programs, e.g., in the eight counties in the HGB area. The program is income-based, with a monetary assistance of $3,000 for cars or trucks and $3,500 for hybrids. The replacement vehicle must meet certain criteria, including being a 2008 or newer model year (*6*).

The Houston-Galveston Area Council (HGAC) Clean Cities/Clean Vehicles program conducted a review of cost and effectiveness of the federal Congestion Mitigation Air Quality (CMAQ) program and determined that the most cost effective use of funds for emissions reductions is to target heavy-duty vehicles (*7*). However, only one program, the Emission Reduction Incentive Grants (ERIG), has been found to target HDDVs. ERIG is part of the Texas Commission on Environmental Quality's (TCEQ) Texas Emissions Reduction Plan (TERP) program and makes grants available to public or private owners of on-road heavy-duty vehicles, construction equipment, marine vessels, locomotives, and stationary equipment in NA areas that wish to upgrade or replace them with newer, cleaner units. The grants offset the incremental costs associated with reducing emissions of NOx from high emitting internal combustion engines (*8*). The program has resulted in over 1000 diesel engine and/or vehicle replacements and is a potential source for vehicle recruitment.

Mobile Source Air Toxics

HDDV emissions also include an array of known or suspected carcinogenic compounds, known collectively as MSAT. MSAT have become increasingly important in recent years as states and air quality districts are now required to address them along with other criteria pollutants. The typical emissions control strategies (tailpipe emissions control, reformulated gasoline, ultra-low sulfur diesel [ULSD]), that target primary pollutants such as NOx also help control air toxic emissions. The EPA has identified 21 toxics, and MOVES provides the option for calculating four major air toxic compounds: benzene; 1,3-butadiene; formaldehyde; and acetaldehyde (*9*). Figure 2 shows a subset of the 21 MSAT compounds identified by EPA and the relationship between the compounds in the subset. The study focused on measuring aldehydes, formaldehyde, and acetaldehyde.

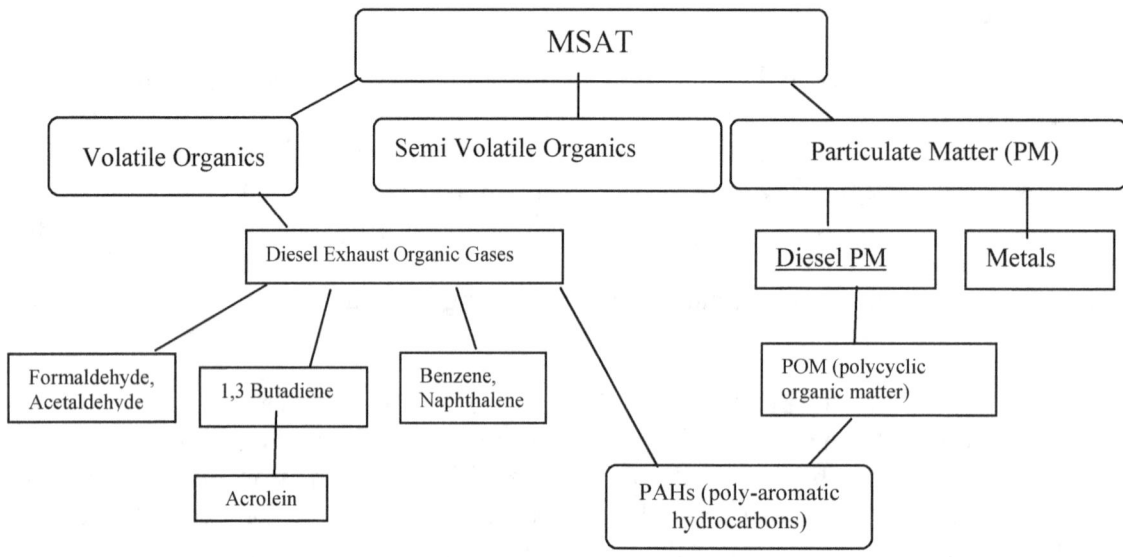

Figure 2. Relationship between MSAT and Diesel-Related Compounds.

TCEQ has recognized the potentially detrimental effect of air toxics on human health and in collaboration with EPA has commissioned the Houston Exposure to Air Toxics Study (HEATS) that aims to better understand Houston residents' daily exposure to air toxics. This study examines whether the exposure rates differ from ambient monitoring data and investigates the potential health risks and possible risk reduction strategies (*10*).

MOVES

One of the goals of this project was to compare the emissions rates measured in this project with those used in MOVES. It is important to understand how this model has been developed because the study design and the development of the duty cycles are dependent upon the model selected for comparison. The following section provides brief descriptions of the MOVES model.

Prior to MOVES, MOBILE6.2 was the EPA-approved model for all regional emissions analyses, with the exception of California, for ozone precursors (Volatile Organic Compounds [VOC] and NOx), CO, and PM (*11*). In response to the demand for an emissions estimation tool that can accommodate finer-scale analysis than regional-level (currently in MOBILE6), the EPA has developed a new model, MOVES (*12*). Unlike the aggregate approach used for the MOBILE6.2 model, MOVES utilizes a disaggregate measure called Vehicle Specific Power (VSP), which is a combined measure of instantaneous speed, acceleration, road grade, and road load. The emissions associated with any given driving pattern are modeled based on distribution of time spent in VSP bins and speeds. In addition to exhaust emissions, MOVES provides estimates of start, brake wear, tire wear, and extended idling emissions.

Drive cycles representing typical operations at different average speeds for each vehicle type operating on a road are used to translate average speed information into VSP distributions. VSP is calculated on a second-by-second basis for vehicle operations over these drive cycles. MOVES then estimates running exhaust emissions and energy consumption based on the total hours of operations in its 33 operating (VSP) mode bins; each bin represents a range of vehicle speeds and VSP.

MOVES estimates energy consumption and mass emissions of pollutants. Energy consumption estimated by MOVES includes total energy consumption, fossil fuel energy consumption, and petroleum fuel energy consumption. The mass emissions, which can be estimated by MOVES, are total gaseous hydrocarbons (THC), CO, NOx, PM (including PM from fuel sulfur, tire wear, and brake wear), methane (CH_4), nitrous oxide (N_2O), carbon dioxide (CO_2), and the "CO_2-equivalent" greenhouse gas emissions of CO_2 combined with N_2O and CH_4.

Despite the structural flexibility of the MOVES model, which enables users to model different driving patterns, the EPA released the model with only national average driving patterns incorporated, which are mostly the same driving cycles used for the MOBILE6.2 model. To take full advantage of the MOVES features, the users must provide local driving patterns as well as other local input. Currently, the MOVES model contains no local values. It includes only national average values. For this project, the research team used driving pattern information collected during the data collection and testing.

The VSP structure in the MOVES model provides for the flexibility required in this study. Instead of using rigid drive cycles, a set of drive patterns can be utilized to capture emissions during different vehicle operational conditions. The drive patterns developed for this study contain different steady-state (cruise speed), acceleration and deceleration rates, and various idling modes. This made it possible to compare the results from this study directly with existing MOVES rates and provided data that can potentially be used as local default values for the HGB area and Texas.

CHAPTER 2: PROJECT APPROACH AND TEST PROTOCOL

OVERALL APPROACH

The overall approach for this research project involved the execution of seven tasks over a three-year period. Figure 3 shows a flow diagram of the project and how the various tasks will fit together. The overall approach was mainly to:
- Develop test protocol.
- Select a fleet.
- Screen HEs.
- Select test vehicles (HEs and randomly-selected HDDVs).
- Develop drive cycle with conducting GPS data collection.
- Perform testing (both driving and idling testing).
- Analyze test data and compare test data with MOVES estimates.

For this project, a three-year duration was necessary to allow enough time to test an adequate number of HDDVs (Class 4, 6, and 8) and to provide an opportunity to use the knowledge gained during the first year to make the second and third year's testing even more effective in terms of test methodology and protocols to be followed.

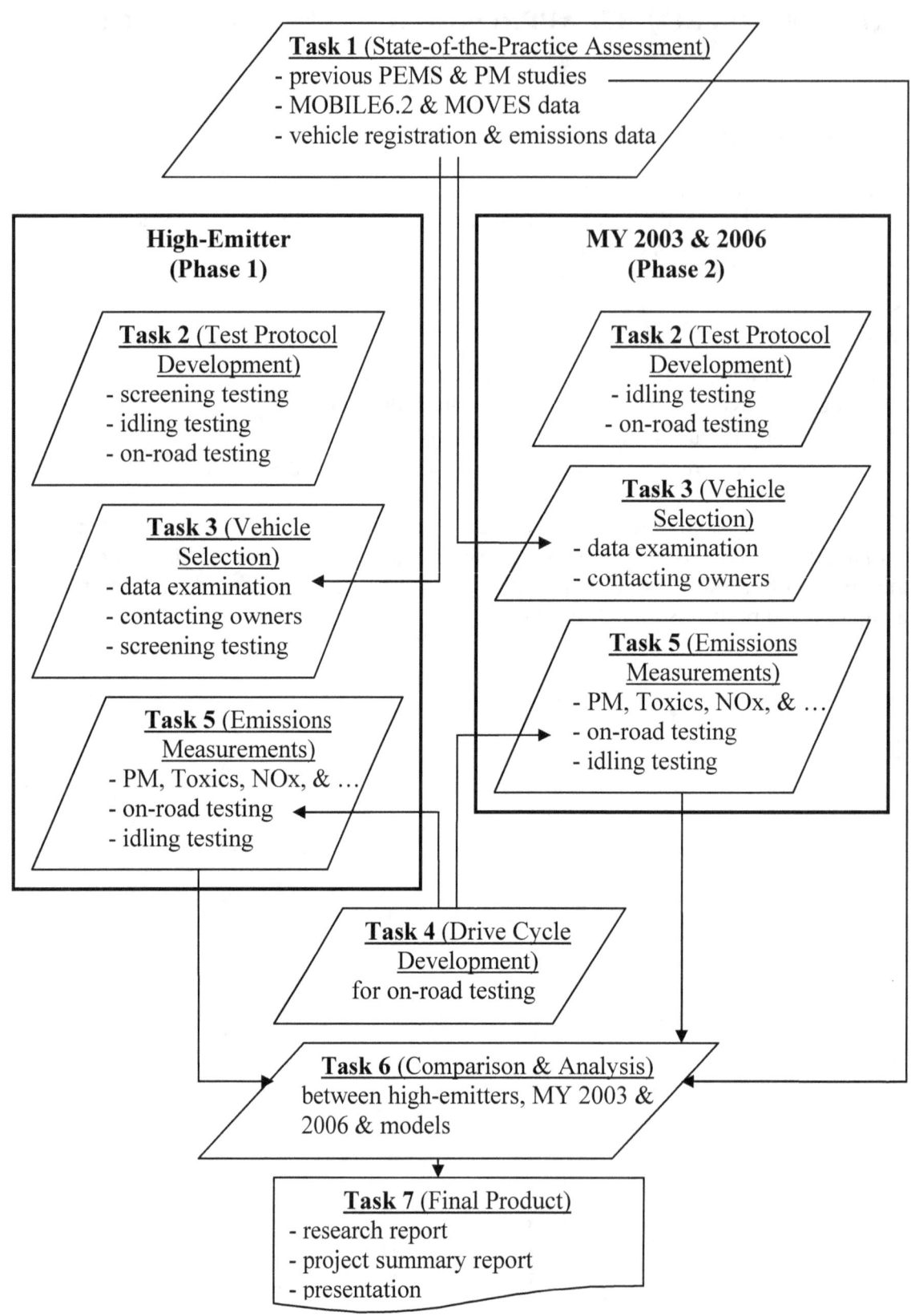

Figure 3. Task Flow Chart.

The project was conducted in two phases. Phase 1 was executed during years one and two and Phase 2 during years two and three. Phase 1 involved background research, development of the detailed test protocols, performing screening testing, and selection of the high emitting vehicles. Phase 2 of the project involved the testing of both high emitting vehicles and random class 8 vehicles from model years (MYs) 2003 and 2006 HDDVs. A total of 30 vehicles were selected for testing, including 12 random samples from MYs 2003 and 2006, and six vehicles from each category, which were selected due to results obtained during the screening process. More details for vehicle selection are described in the next chapter.

All testing of the vehicles was conducted at TTI's Environmental and Emissions Research Facility (EERF) and runway facilities, located at Texas A&M's Riverside campus. Each vehicle arrived at the EERF one day prior to testing to be outfitted with all the necessary test equipment. Figure 4 shows one of the class 4 vehicles with instruments ready for testing.

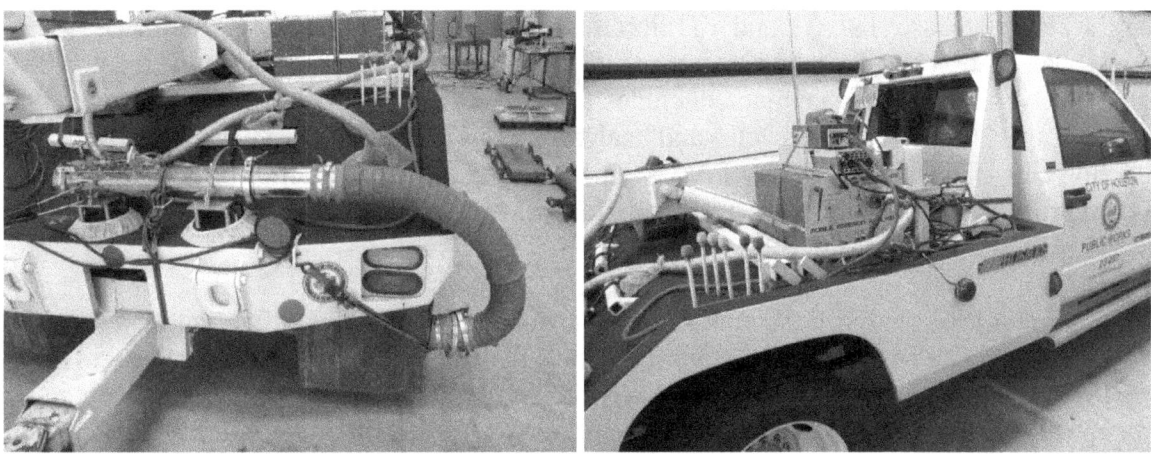

Figure 4. Class 4 Vehicle Instrumented for Testing.

For each vehicle, driving testing was conducted first, then, idling testing followed. The drive cycle chosen for the driving testing is described in the following section in detail. During the driving testing, the chamber was prepared for the idling testing. The idling testing condition was chosen to represent a typical weather condition that the vehicle would see during the year in the HGB area. The selected condition for the idling testing was 86°F with 60 percent relative humidity (RH). Once the driving testing was completed, the test vehicle was brought inside the test chamber and allowed to idle for approximately one hour. During the idling testing, PM filter and MSAT cartridge samples were taken in addition to the emissions measurements with using portable emissions measurement system (PEMS). Each sample was collected for 30 minutes. Each sample was controlled by separate pumps, one for the cartridge and one for the filter sample. These pumps assured that an accurate reading of the volume of each sample was taken. After the idling testing was complete the testing equipment was removed from the vehicle and the next vehicle was prepared for the next testing. This testing procedure was followed to test all of the 30 vehicles.

DRIVE CYCLE DEVELOPMENT

Using a drive cycle rather than just collecting data on a normal workday operation (i.e., when the vehicle is in use) has two main advantages. First, use of drive cycles enables tests to be repeated, which improves the quality of the data. Second, drive cycles can incorporate more operation modes (2), even the high-emissions events that may not typically occur. For example, if data were to be collected during a normal workday, then data reflecting high engine loads would not be adequately collected. For this study, a typical drive cycle is augmented with a high acceleration at high speed portion where vehicles are forced to work at maximum load. This enables collection of data for such high-emissions events.

In order to develop simplified and repeatable drive cycles that also could be used for all possible operation conditions of the tested vehicles, the research team conducted search of published and unpublished materials using personal contacts, databases such as the Transportation Research Board's TRIS database, TxDOT and TTI libraries, EPA and California Air Resources Board (CARB) databases, and general web searches to obtain information on diesel-powered non-road equipment. The research team then concluded that two different types of drive cycles would be needed for this study: test drive cycles and analysis drive cycles. Test drive cycles were used for collecting emissions data based on EPA MOVES' general modeling framework. A MOVES-based approach was found to be more suitable for PEMS testing and was therefore used for this study.

After the researchers analyzed the emissions data and determined the emission rates, a series of analysis drive cycles were needed to determine the emissions impact of a vehicle class. These drive cycles were speed profiles of the real-world operation of the target vehicles collected using global positioning system (GPS) devices. After selecting vehicles to be tested, the research team collected GPS data regarding the operational characteristics of the selected vehicles. These data were then used to determine the distance-based emissions rates and overall emissions of the selected vehicle classes.

The research team used a series of drive cycles for driving testing. All test runs for each vehicle from the same vehicle class followed the same set of drive cycles. Each vehicle was tested repeatedly following the drive cycles at least 3 times (up to 10 times). For this study, a typical drive cycle is augmented with a high acceleration at high speed portion where vehicles are forced to work at maximum load. This enables collection of data for such high-emissions events.

Drive Cycles for Emissions Testing

This project utilized some of the concepts and methodologies used in the EPA's MOVES model to analyze emissions data. The MOVES model has been recently released and will officially replace EPA's MOBILE 6.2 model for air quality planning, emissions inventories, and regulatory efforts in 2012. By using the MOVES methodology framework to analyze data, this project was able to analyze the operational characteristics of emissions data according to an established analysis protocol. The local data needs for MOVES are also significant, and the results from this study can be used to enhance localized inputs into the MOVES model for the HGB area.

The MOVES model characterizes in-use emissions by using second-by-second emissions rates that account for a vehicle's operating modes. This enables the model to provide a finer scale of analysis at a local level than that provided by the MOBILE6.2 model. MOVES incorporates VSP to capture modal emissions. EPA defines VSP as "power per unit mass of the source" and is characterized in the VSP equation below (*13, 14*). VSP accounts for the forces a vehicle must overcome when operating on the road, including acceleration, road grade, tire rolling resistance, and aerodynamic drag. For example, fast accelerations or driving up a steep hill would have a higher VSP bin rather than coasting downhill.

Equation 1: VSP Calculation

$$VSP = \frac{A \times u + B \times u^2 + C \times u^3 + M \times u \times a}{M}$$

where:
- A = A rolling resistance term
- u = Instantaneous speed of vehicle
- B = Rotating resistance term
- C = Drag term
- M = The vehicle's mass
- a = Instantaneous acceleration of vehicle

VSP is normalized by mass, and then operating mode bins are determined from VSP and instantaneous speed. There are 23 operating bins for running emissions (i.e., when vehicles are moving or idling at hot-stabilized conditions). Table 1 shows the MOVES' operating mode bins. A vehicle operating over a drive cycle spends different times in different bins depending on the operation.

Table 1. MOVES Operating Mode Bin Definitions for Running Emissions.

Braking (Bin 0)			
Idle (Bin 1)			
	Instantaneous Speed (mph)		
Instantaneous VSP (kW/tonne)	0–25	25–50	> 50
< 0	Bin 11	Bin 21	
0 to 3	Bin 12	Bin 22	
3 to 6	Bin 13	Bin 23	
6 to 9	Bin 14	Bin 24	
9 to 12	Bin 15	Bin 25	
12 and greater	Bin 16		
12 to 18		Bin 27	Bin 37
18 to 24		Bin 28	Bin 38
24 to 30		Bin 29	Bin 39
30 and greater		Bin 30	Bin 40
6 to 12			Bin 35
< 6			Bin 33

The VSP-based approach provides the flexibility required at meso- and micro-scales of analysis (*15*). MOVES' disaggregated methodology is expected to estimate emissions more precisely based on available local activity data. This approach is a better fit for incorporating second-by-second operational data provided by in-use emissions testing.

As expressed by the VSP equation above, the VSP of a specific vehicle is a function of instantaneous speed and acceleration. The impact of road grade is modeled as the effective gravitational acceleration in parallel to a vehicle's moving direction and therefore is added to vehicle acceleration. This means that bins corresponding to high-VSP values can be achieved by either driving at high acceleration rates or accelerating on high-grade roads.

To obtain enough observations in each modal bin, the driving events must contain different steady state (cruise speed) conditions as well as different levels of acceleration and deceleration rates. Note that unlike laboratory testing, having pre-specified second-by-second driving tracks for on-road data collection is impractical; however, a series of specific driving events that follow pre-defined patterns can easily cover a broad-range of vehicle operating modes. Younglove et al. suggested building such driving events for on-road emissions testing (*16*).

After careful examination of the EPA's MOVES model modal bins (Table 1) and the testing area (the runway facilities in Bryan, TX), the research team designed a set of driving events that consisted of the following elements:

- Maximum acceleration to 3 different speed levels: *low* (15 mph), *medium* (40 mph), and *high* (minimum 60 but not higher than 70 mph) in steps; driving for at least 5 seconds after achieving each speed level.

- Slow acceleration to the *high* speed (as defined above) and maintain the speed for at least 5 seconds.

- Normal acceleration to the *high* speed (as defined above) and maintain the speed for at least 5 seconds.

The flexible structure of the above drive cycles enabled the research team to make necessary changes to the way the drive cycles were executed while retaining the core elements. For example, if the available track length was limited, the maximum acceleration element was broken into two sections with one section containing acceleration to 15 mph and the other section covering the other speeds. These basic drive cycle elements were applied to all three classes of test vehicles with few modifications in execution based on the circumstances in the field.

Test vehicles repeated each event on a level road/track for a minimum 3 (up to 10) repetitions. A period of at least 10 seconds of idling was included in between each run to allow the engine to stabilize to unloaded conditions and, therefore, to minimize any possible effects from the previous run. A normal deceleration was considered for all events.

Drive Cycles for Emissions Analysis

The GPS data were collected using City of Houston's vehicles in February and March 2010. The research team collected the GPS data for two days for each vehicle. Vehicles selected for testing had GPS units installed in the morning before they left the shop for their daily work. After returning to the shop in the evening, the GPS units remained on the vehicles overnight and were kept until the completion of a second day's work. The GPS units were removed on the vehicles' return to the shop. Thus GPS data for two full days were collected for each vehicle. A total of 18 vehicles were selected for the GPS data collection. Table 2 summarizes the number of vehicles in each class used for the GPS data collection.

Table 2. Summary of Vehicles in GPS Data Collection.

Vehicle Class	Number of Vehicles Tested
HDDV 4	5
HDDV 6	5
HDDV 8	8

Figure 5 shows an example of daily speed profiles. The speed profile corresponds to a single day's operation of a HDDV 8 vehicle. The research team developed similar profiles for the other vehicle classes. The collected speed profile data were used to determine the distance-based emissions rates and to characterize the overall emissions of the selected vehicle classes in the study area as well as to make comparisons with MOVES outputs. Figure 5 and Figure 6 demonstrate this process graphically.

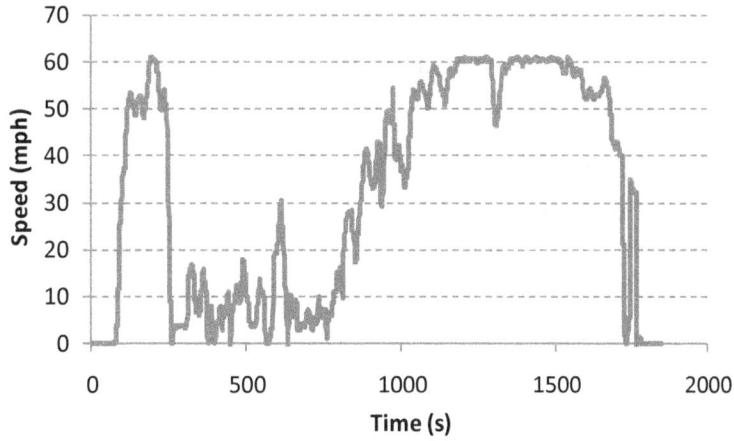

Figure 5. Sample Daily Speed Profile for a HDDV 8 Vehicle.

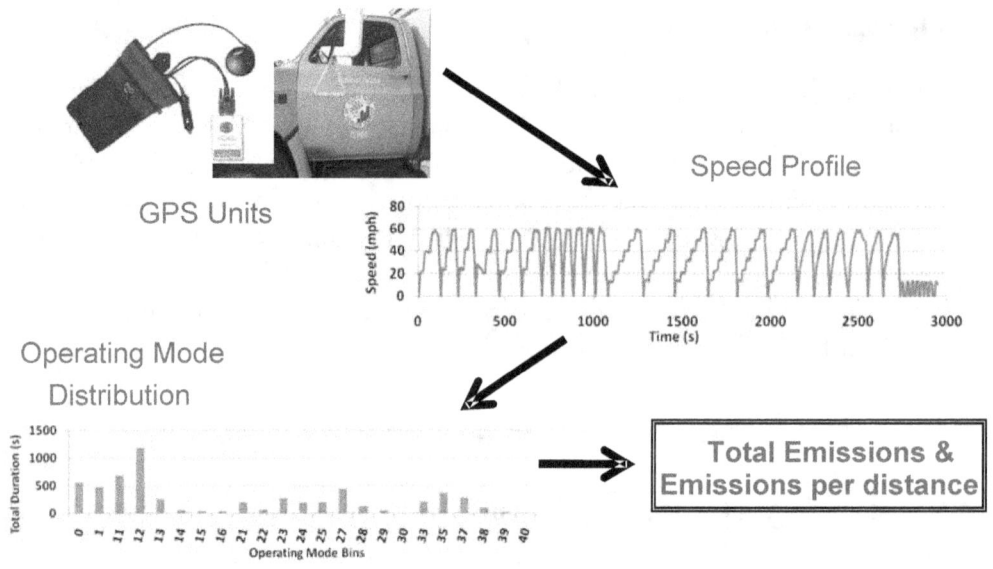

Figure 6. Using Real-World GPS Data for Emissions Analysis.

TEST FACILITIES AND EQUIPMENT

TTI Environmental and Emissions Research Facility

All the testing for this project was conducted at TTI's EERF. The EERF is a test facility containing climate controlled test chamber located at Texas A&M University's (TAMU's) Riverside campus. The environmentally-controlled test chamber has dimensions of 75×23×22 feet. The chamber is the largest vehicle test chamber in the nation, capable of producing a temperature range of −40°C to 55°C. The chamber also has humidity controls capable of having a RH of 70 percent at 40°C. Other equipment inside the chamber includes a solar light array to simulate sunlight and a wind simulator capable of producing wind speeds of up to 25 MPH. In addition to conducting the idle testing inside the test chamber the driving testing was conducted on the runways adjacent to the facility. The TAMU Riverside campus includes a set of runways. The runways allow for all the driving conditions required for the proposed drive cycles. Figure 7 shows one of the Class 6 vehicles being tested inside the EERF chamber.

Figure 7. City of Houston Vehicle inside EERF Chamber.

Test Equipment

The EPA rule (*1*), "In-Use Testing Program for Heavy-Duty Diesel Engines and Vehicles," governs the way that PEMS testing must be performed. For this rule, the EPA's emissions testing procedure 40 CFR Part 1065 describes the PEMS testing procedure for gaseous sampling including NOx, CO, HC, CO_2 in a high level of detail and specifies the instruments required for these tests; for example, a flame ionization detector (FID) needs to be used to measure HC emissions. Also, it specifies that a flow meter must meet certain specifications in order to be used for measuring exhaust flow. TTI's PEMS equipment complies with all the specifications of the EPA rules. For MSAT and PM, however, no official PEMS testing procedure has been approved by the EPA yet.

The research team used two PEMS simultaneously during the driving testing—one to collect gaseous emissions NOx, HC, CO, and CO_2 and the other one to collect PM emissions. TTI's SEMTECH-DS unit was used along with TTI's electronic vehicle exhaust flow meters (EFMs) for the gaseous emissions. TTI's Axion was used to measure PM emissions. For DPM and other toxics characterization, the research team collected integrated samples from dilution systems while the HDDVs performed their idling testing. In addition, PM mass was monitored using a continuous PM mass monitor (Dekati Mass Monitor, DMM) during the idling tests. TTI's micro dilution sampling system (MSS) was used to collect both filter and cartridge samples of the vehicles during the idling tests. Collected filter and cartridge samples were sent to Oak Ridge National Laboratory for PM and MSAT analysis. The following is a brief description of the PEMS, MSS, and DMM.

SEMTECH-DS

The SEMTECH-DS system includes a set of gas analyzers, an engine diagnostic scanner, a GPS, an EFM, and embedded software. The gas analyzers measure the concentrations of NOx (NO and NO_2), HC, CO, CO_2, and oxygen (O_2) in the vehicle exhaust. The SEMTECH-DS uses the Garmin International, Inc. GPS receiver model GPS 16 HVS to track the route, elevation, and ground speed of the vehicle on a second-by-second basis. TTI's SEMTECH-DS uses the SEMTECH EFM to measure the vehicle exhaust flow. Its post-processor application software uses this exhaust mass flow information to calculate exhaust mass emissions for all measured exhaust gases. Figure 8 shows a picture of the SEMTECH-DS and EFM installed on a test vehicle prior to driving testing.

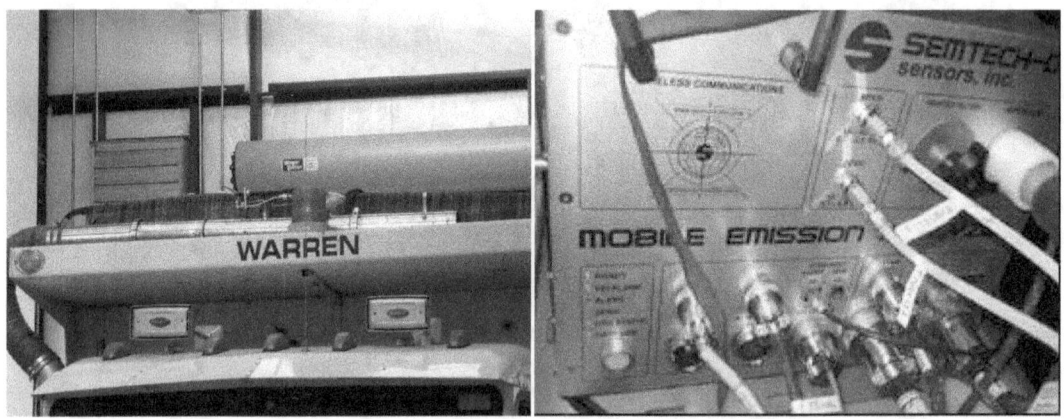

Figure 8. SEMTECH EFM Installed on Test Vehicle (Left) and SEMTECH-DS (Right).

Axion

The PEMS used to collect PM was an Axion system manufactured by Clean Air Technologies International, Inc. The Axion system consists of gas analyzers, a PM measurement system, an engine diagnostic scanner, a GPS, and an on-board computer. For this study only the PM measurement system was used. The PM measurement capability includes a laser light scattering detector and a sample conditioning system. The PM concentrations are converted to PM mass emissions using concentration rates measured by the Axion and the exhaust flow rates collected by the SEMTECH EFM. Figure 9 shows a picture of the Axion system along with the SEMTECH-DS installed on a test vehicle prior to testing.

Figure 9. CATI Axion System along with the SEMTECH-DS Installed on Test Vehicle prior to Testing.

Microdilution Sampling System

PM and other MSAT will be sampled using a MSS similar to the one used in previous TTI studies on idling trucks (*17, 18*). The exhaust was transferred through a heated line to the microdilution tunnel from a probe in the outlet of the SEMTECH EFM. For each idle condition, PM and other toxics were collected on filters and in a Solid Phase Extraction (SPE) cartridge. The exposed filters and cartridges were then sent to Oak Ridge National Laboratory for analysis. Also, PM mass was monitored continuously by using a DMM from the diluted exhaust. To determine dilution ratio, NOx measurements were made for both the raw and diluted exhaust. Figure 10 shows a picture of TTI's MSS.

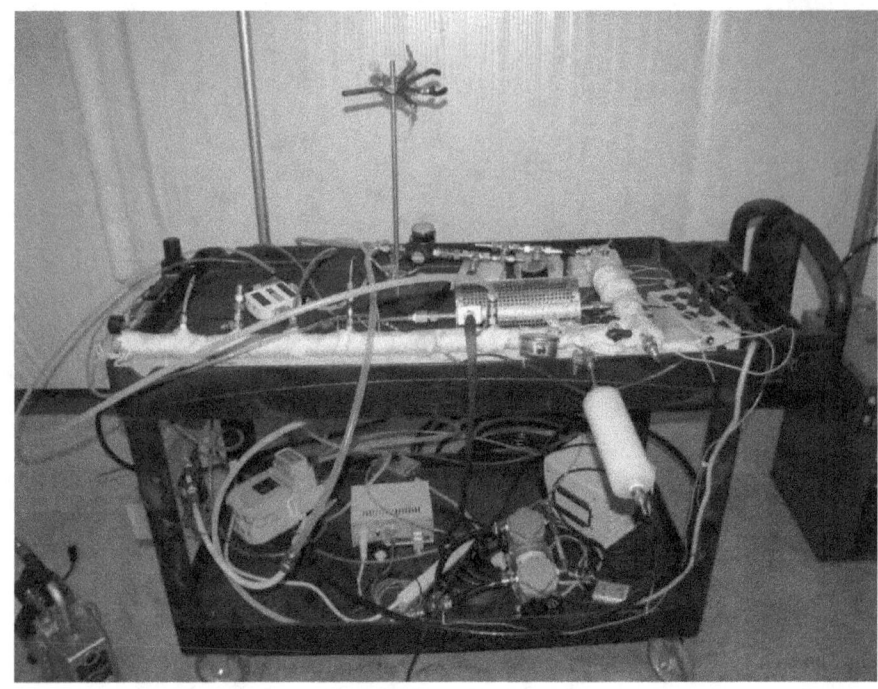

Figure 10. Microdilution Sampling System.

Dekati Mass Monitor

The DMM 230-A is a real time measuring device for particulate matter for automotive testing. The DMM can be used for both gasoline and diesel engines. The DMM provides second-by-second analysis of the PM concentrations from the vehicle exhaust, providing both total mass measurements and the size of the particles, from 0–1.5 μm. The DMM was used in conjunction with the MSS to measure the diluted exhaust directly out of the engine. Figure 11 shows TTI's DMM system.

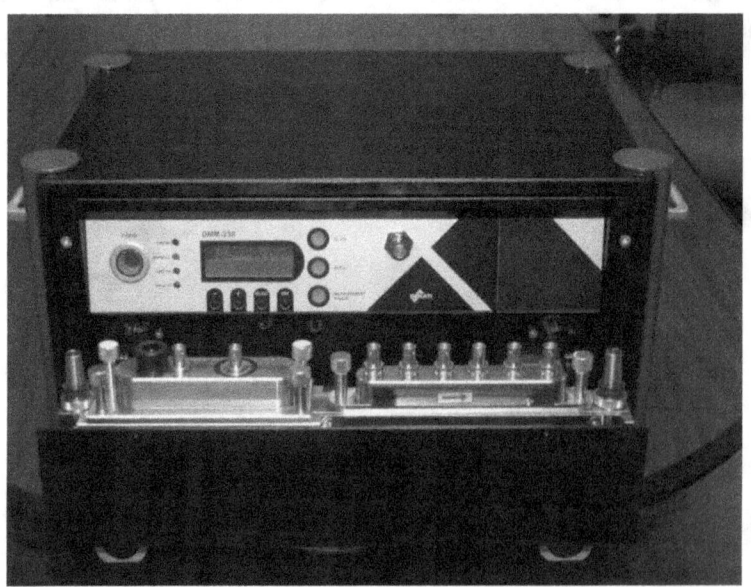

Figure 11. Dekati Mass Monitor.

CHAPTER 3: TEST VEHICLE SELECTION

It is well known that HDDVs, of which gross vehicle weight rating (GVWR)s are over 8500 lb, emit higher amounts of PM and NOx compared to their gasoline-powered counterparts. According to the latest estimations of the California Air Resources Board (CARB), HDDVs are responsible for 40 percent of NOx and 32 percent of PM emitted by diesel mobile sources in the state of California (*19*).

TARGET VEHICLES

Among HDDVs, Classes 4, 6, and 8 of MYs 2003 and 2006 were of special interest to TxDOT and were selected for testing under this project because they were highly prevalent groups and have the highest potential to provide a representative sample of the HDDV population to indicate real-world HDDV emissions levels. HDDV registration data for the eight-county HGB area, provided by TxDOT as of 7/31/09, showed that there were a total of 67,822 HDDVs, and more than 20 percent of HDDVs (14,668) belong to these three classes and with MYs 2003 and 2006. Table 3 shows a statistical overview of the registration data for the HGB area.

Table 3. HGB Area – Select HDDV Classes.

Model Year*	Registration Statistics	EPA Vehicle Class*		
		HDDV4 (3981)	HDDV6 (8284)	HDDV 8B (2403)
2006 (7555)	Number	441	727	308
	% of MY Registrations	5.84%	9.62%	4.08%
	% of Vehicle Class registrations	11.08%	8.78%	12.82%
2003 (3799)	Number	248	364	131
	% of MY Registrations	6.53%	9.58%	3.45%
	% of Vehicle Class registrations	6.23%	4.39%	5.45%

* Classification Totals in ( )

HEs

Because HEs contribute disproportionately to overall HDDV emissions, as stated and shown in Chapter 1, HEs were also selected to provide their emissions impacts. A quantitative definition of an HE has not been established, so the term "HE" is employed qualitatively. The current EPA PM emission standard for MY 2007 or newer heavy-duty diesel engines is 0.01 g/bhp·hr, and, for MY 1994–2006, it is 0.1 g/bhp·hr (*20*). Today, a high PM emitter is considered to be any vehicle that emits 2.5 times or greater PM than the current regulatory standard. Numerous studies have found that, on average, those with visible smoke emit as much as 10 times or more PM than standard non-smoking vehicles (*21*). A more thorough discussion of the process used for selecting the HEs for testing in this project is shown in the following section.

VEHICLE SELECTION PROCESS

Fleet Selection

The research team began its search for a fleet of vehicles to be tested. Finding an available fleet of HDDVs turned out to be a difficult and time consuming task. The research team examined many different options to recruit a test fleet. The options included:
- A large fleet of vehicles owned by one group that could provide all necessary vehicles for testing.
- Renting vehicles one at a time from multiple vendors.
- Testing vehicles that were replaced via the Congestion Mitigation Air Quality program.

The research team checked more than a dozen possible sources of high emitting vehicles. After examining the available options and discussing with TxDOT, the research team determined that the best source was the City of Houston fleet. In order to utilize the COH fleet, it was necessary that TTI would obtain a legal agreement (or contract) from COH. Once the contract was finalized the research team was given access to the COH fleet vehicles as well as COH drivers. Under the agreement the COH provided the research team access to their fleet for opacity measurement and GPS data collection. Furthermore, selected vehicles were available for emissions testing at the EERF in Bryan. COH fleet drivers delivered each of the selected vehicles to the EERF and operated them during the emissions testing.

HEs Selection

The process of selecting the HEs began with opacity testing of potential candidates. Opacity testing is a measurement of the amount of light that cannot pass through the vehicle exhaust, expressed as a percentage. The opacity reading is an indication of the amount of emissions from a vehicle; the higher the opacity the higher the expected exhausts emissions. The equipment used for opacity testing was a Wager 7500 smoke meter. The smoke meter is in compliance with the instrument specification in Society of Automotive Engineers (SAE) J1667 test procedure (*22*).

SAE J1667, *Snap Acceleration Smoke Test Procedure for Heavy-Duty Diesel Powered Vehicles*, is the standard that covers exhaust smoke measurements. The SAE J1667 standard outlines the procedures for conducting snap acceleration smoke testing on HDDVs. The research team followed the procedures to obtain opacity readings of candidate vehicles for selecting potential HEs for further emissions testing. According to the procedure a snap acceleration test is performed in four steps.

1. The vehicle operator throttles the vehicles to the fully open position as quickly as possible.
2. This position is maintained by the operator until the engine reaches its fill governed speed, plus approximately 4 seconds.
3. After this time the throttle is released and the engine is allowed to return to a low idle condition.

4. After the engine reaches the low idle condition it must be left in that condition for a minimum of 5 and maximum of 45 seconds before another snap acceleration test cycle is conducted.

More details of the test can be obtained from the SAEJ1667.

The opacity screening tests were conducted between December 4, 2009, and January 26, 2010. Tests were conducted by Texas Southern University (TSU) with guidance from TTI. For each round of tests the TSU team traveled to a COH facility and performed opacity tests on multiple vehicles. The team conducted screen tests of 91 vehicles across 7 classes in 8 rounds of tests. Figure 12 shows pictures taken while the team was conducting screening tests of Class 6 vehicles from the COH fleet.

Figure 12. Pictures of Opacity Testing on COH Fleet Vehicles.

After the opacity screening tests were complete, the research team used the collected opacity data to select HEs. It should be noted that the original plan to test Class 2b vehicles in the COH fleet was modified to test Class 4 vehicles instead because the number of available Class 2b vehicles was too small and the Class 4 trucks in the COH fleet have the same engine as Class 2b trucks. The only differences between Classes 2b and 4 trucks were in the suspension, axle, and frame length. Consequently, there was almost no expected difference in terms of emissions between the Class 4 and Class 2b vehicles. The similarity of engine and emissions classification is also acknowledged by EPA as all the diesel pickup trucks (both Classes 2b and 4) fall into the same source type (32, light commercial trucks) in EPA's newest emissions model, MOVES (22).

From the screening tests, the research team found that all of the opacity readings of Class 6 vehicles were very low. The highest opacity reading was 17 percent, which was lower than those of vehicles in other classes that were considered as HEs; the average was 14 percent. Based on the test results, the research team decided to classify the tested Class 6 vehicles as normal emitting vehicles not HEs.

After analyzing all of the collected opacity reading data the team determined that the top 6 opacity readings for vehicles in Classes 4, 6, and 8b would be used for in-use emissions testing. Table 4 shows the vehicle information and opacity results from the selected vehicles. Figure 13 also demonstrates the cumulative distribution of the observed opacity reading for the three selected vehicle classes.

Table 4. Selected Vehicles Opacity Results.

Class 4		Class 6		Class 8b	
MY - Make	Opacity Reading (%)	MY - Make	Opacity Reading (%)	MY - Make	Opacity Reading (%)
1999 Chevy	72	1993 Chevy	17	1995 Ford	40
1999 Chevy	64	2001 International	17	2003 Sterling	29
1999 Chevy	54	2000 International	13	1993 Ford	24
1994 Chevy	43	2001 International	12	1994 Ford	24
1999 Chevy	35	2000 Volvo	12	1996 Ford	22
1999 Chevy	32	1999 International	11	2003 Sterling	20

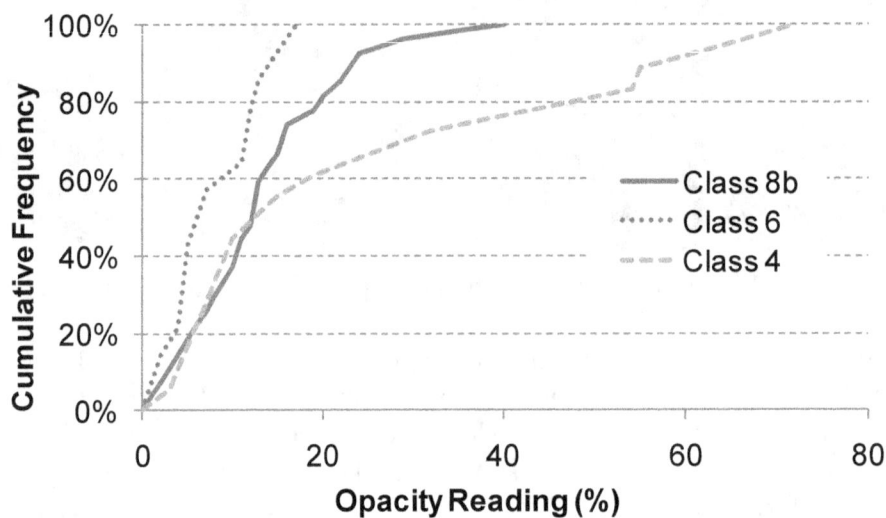

Figure 13. Cumulative Frequencies of Opacity Readings.

In addition to the 18 vehicles, the 12 non-HE Class 8 vehicles of MYs 2003 and 2006 were also selected randomly from the COH fleet for the emissions testing. That is, for the emissions testing, a total of 30 vehicles were selected—18 vehicles by opacity tests and 12 vehicles by random selection. The emissions testing with the selected vehicles began on April 27, 2010.

CHAPTER 4: TESTING RESULTS AND ANALYSES

Emissions testing results were measured in two different parts of testing: driving testing and idling testing. During the driving testing, test vehicles followed the developed drive cycles on runways adjacent to the EERF. Using two PEMS, gaseous and PM emissions were measured during the driving testing with various driving conditions, that is, different speeds, accelerations, and engine loads. Idling testing was conducted inside EERF at one test condition (86°F with 60 percent RH) at one low engine speed. In addition to gaseous and PM PEMS measurements, PM and MSAT samples and PM mass monitoring were also taken during the idling testing as described in Chapter 2. Due to the different nature of driving testing (various driving condition) and idling testing (only one idling condition), results of each testing were analyzed differently and reported in different subsections below. However, both testing results contain comparisons between classes, vehicles, and measured vs. MOVES estimates, respectively.

In addition, based on both driving and idling testing results, possible emissions benefits by replacing HEs were discussed in the subsection, "Emissions Reduction Analysis Benefit." Also, relationships of different PM measurement methodologies, which can be used to identify HEs, are discussed in the last subsection.

During this project, 30 vehicles were tested in the 4 different categories: Class 8 randomly selected vehicles, Class 8 HEs, Class 6 vehicles, and Class 4 HEs. Table 5 through Table 8 show details for each vehicle that was tested, including make, MY, and mileage at the beginning of testing.

Table 5. Tested Class 8 HEs.

Class 8 HEs Vehicle Information				
Unit Number	COH Vehicle Number	Make	Model Year	Beginning Mileage
8H-1	21021	Ford	1993	NA (had been replaced)
8H-2	21832	Ford	1994	152,494
8H-3	23718	Ford	1995	35,222
8H-4	25092	Ford	1996	127,395
8H-5	33315	Sterling	2003	53,602
8H-6	33309	Sterling	2003	96,088

Table 6. Tested Class 8 Randomly Selected Vehicles.

Class 8 Randomly Selected Vehicle Information				
Unit Number	COH Vehicle Number	Make	Model Year	Beginning Mileage
8-1	33405	Sterling	2003	66,567
8-2	33907	Peterbilt	2003	37,856
8-3	33903	Peterbilt	2003	93,493
8-4	33400	Peterbilt	2003	65,055
8-5	33404	Peterbilt	2003	68,045
8-6	33905	Peterbilt	2003	80,455
8-7	33406	Peterbilt	2003	69,565
8-8	33906	Peterbilt	2003	92,077
8-9	33403	Peterbilt	2003	82,362
8-10	33980	Peterbilt	2006	101,656
8-11	33988	Peterbilt	2006	60,233
8-12	33989	Peterbilt	2006	53,444

Table 7. Tested Class 6 Vehicles.

Class 6 Vehicle Information				
Unit Number	COH Vehicle Number	Make	Model Year	Beginning Mileage
6-1	21907	Chevy	1994	99,595
6-2	29332	International	1999	68,221
6-3	30624	International	2000	47,299
6-4	30424	International	2000	48,864
6-5	30625	International	2001	66,540
6-6	31514	Volvo	2002	90,483

Table 8. Tested Class 4 Vehicles.

Class 4 HEs Vehicle Information				
Unit Number	COH Vehicle Number	Make	Model Year	Beginning Mileage
4H-1	23497	Chevy	1994	114,093
4H-2	29408	Chevy	1999	108,543
4H-3	29540	Chevy	1999	68,087
4H-4	29417	Chevy	1999	100,332
4H-5	29497	Chevy	1999	80,315
4H-6	29415	Chevy	1999	118,483

DRIVING TESTING RESULTS

During the driving testing, second-by-second PEMS data included the following information:
- Engine parameters (if recorded), from the on-board diagnostic system, such as engine speed, throttle position, and engine load for data quality checking.

- Second-by-second vehicle speed from the GPS in mph.
- Emissions mass rates in grams per second (g/s).

The collected second-by-second emissions data were carefully aligned with the collected instantaneous speeds obtained from the GPS data and the engine parameters. For the driving emissions analysis, a linear smoothing was applied to speed data to cancel out noise and fine-scale changes due to GPS accuracy limitations and other factors. The VSP value corresponding to each instance was calculated based on the equation on page 13 utilizing MOVES2010a default parameter values. Table 9 lists the values of parameters A, B, and C used for VSP calculations for each vehicle class. These values were obtained from the MOVES source database.

Table 9. MOVES' Vehicle Parameters Values.

	MOVES Vehicle Type (MOBILE Vehicle Class)	Rolling Term A	Rotating Term B	Drag Term C	Vehicle Mass (tonne)
Combination Short-Haul Truck	Type 61 (HDDV 8)	1.96354	0	0.00403054	29.3275
Single Unit Short-Haul Trucks	Type 52 (HDDV 6)	0.56193	0	0.001603	7.64159
Light Commercial Trucks	Type 32 (HDDV 4)	0.235008	0.003039	0.000748	2.05979

The second-by-second emissions rates were then grouped into operating mode bins according to Table 1. Modal average emissions rates (average emissions rates for each operating bin) for all the pollutants were then estimated for each bin using all the observations that fall into that modal bin. Not all the bins had enough data to determine their corresponding emissions rates. This is usually the case for high-VSP bins for types 52 and 61 vehicles (i.e., Classes 6 and 8 vehicles, respectively), which barely occur during the normal operation of these vehicle classes. For these bins, it was assumed that their emissions rates were equal to the bins that had immediate lower VSP limits in the same speed category. For example, if emissions rates were missing for bin 16, emissions rates of bin 15 were used instead. It was determined that the errors introduced by applying this assumption was minimal because the number of high VSP bins that had a few data during the normal operation were very small compared to lower VSP bins that had a plenty of data during the normal operation.

Modal Emissions

Figure 14, Figure 15, and Figure 16 show the average measured (or observed) and MOVES-estimated emissions rates of all operating modes for a sample vehicle from each vehicle class. Due to an unidentified malfunction of PEMS, emissions data for one of the potential Class 8 HE vehicles (8H-6) could not be validated, so that its data are not reported. As Figure 14 shows, there were no observed data for higher VSP bins, which corresponded to higher engine loads, and therefore they are assumed to be equal to their previous bins in comparison analysis with MOVES estimates. The results for the other vehicles were qualitatively similar, so that they are not presented in this report.

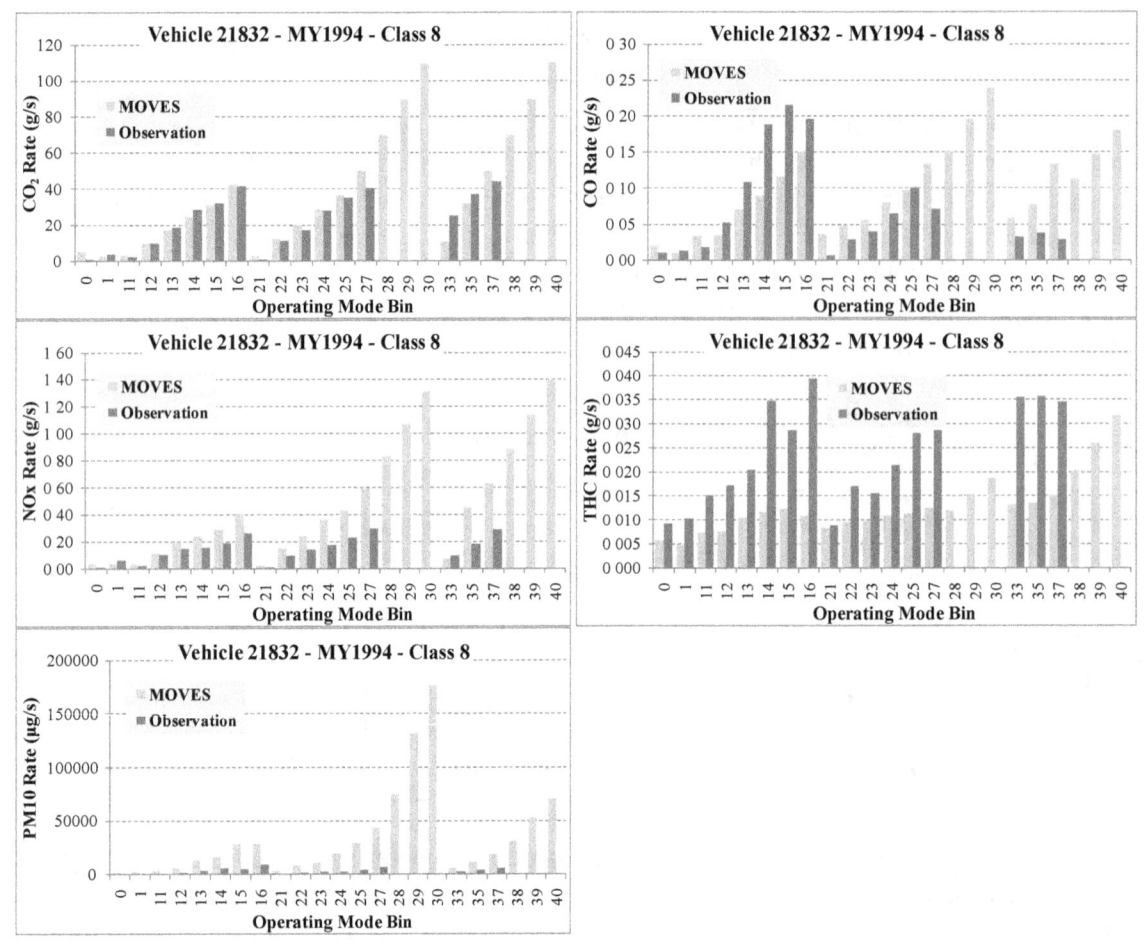

Figure 14. Average Modal Emission Rates for Vehicle 8H-2 (21832).

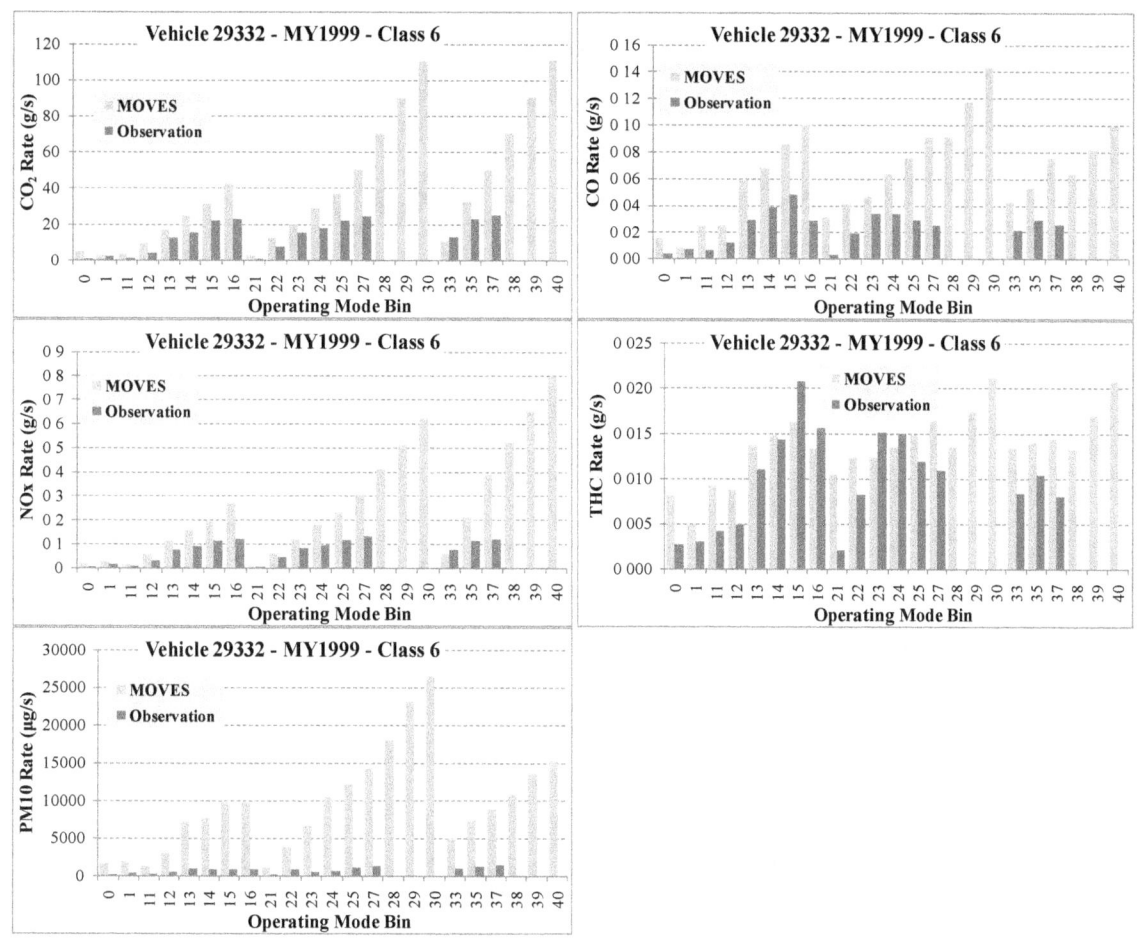

Figure 15. Average Modal Emission Rates for Vehicle 6-2 (29332).

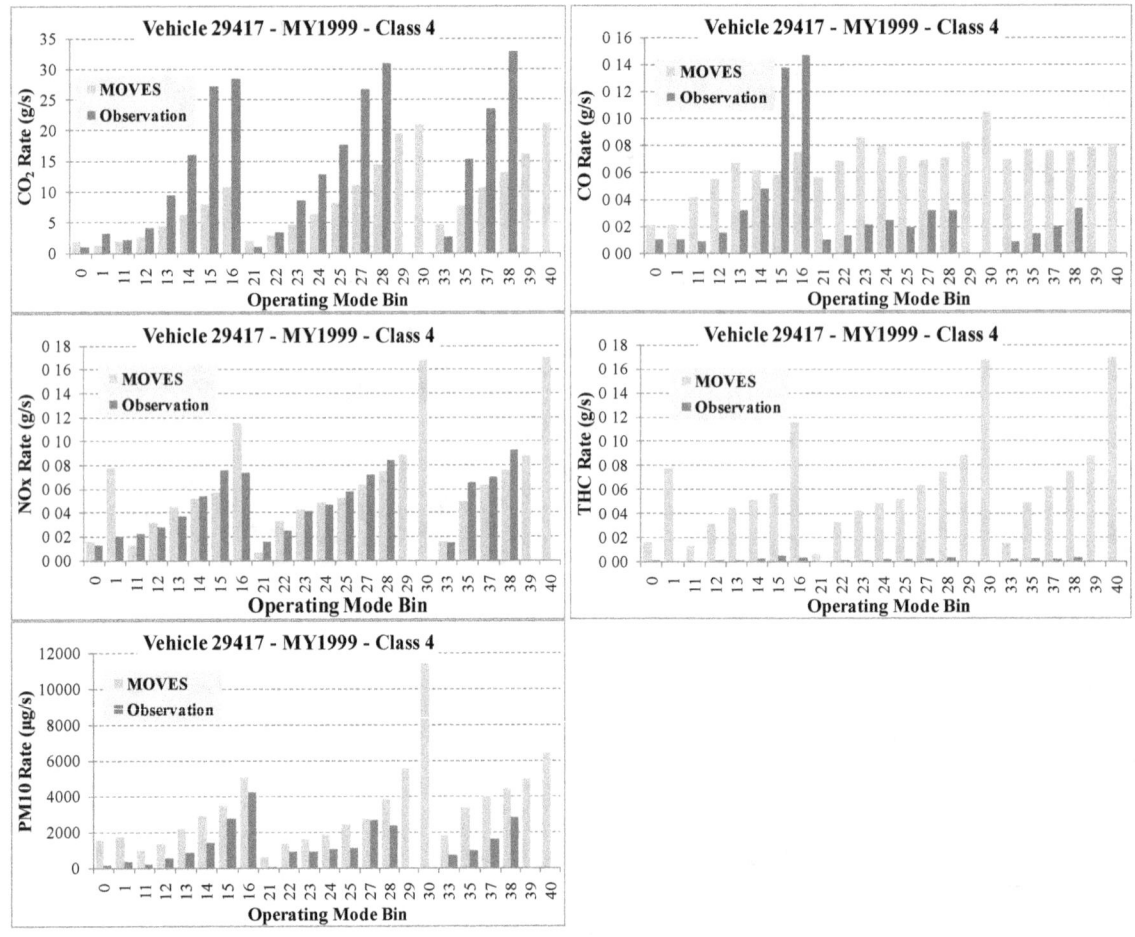

Figure 16. Average Modal Emission Rates for Vehicle 4H-4 (29417).

The emissions rates shown in Figure 14 through Figure 16 also indicate that the modal emissions rates generally increase as the VSP value increases. Exceptions to this trend were found in bins having fewer observations compared to their lower VSP bins having more observations. However, the impact of this issue on the overall emissions results is minimal because very few instances of these high VSP bins occur during normal operations of these vehicle classes.

CO_2 and NOx emissions rates show a consistent trend—higher VSP bins have higher emissions rates. CO_2 formation directly results from engine combustion and therefore is a strong indicator of fuel consumption. Combustion temperature is the main factor determining the NOx formation rate. Higher engine load requires more fuel and therefore produces more CO_2 and heat. Consequent raised engine temperature leads to more NOx formation. Note that high load conditions are very small fractions of the normal operation, and the majority of the normal operation conditions were delivered by low and medium engine loads.

To validate the modal emissions rates, the researchers compared observed total emissions and estimated from the observed average modal rates. The average modal rates were applied to each corresponding instance of the observed data, and the total emissions were calculated for the entire duration of testing. The percentage difference between these estimated values and observed total emissions (reading from PEMS) were calculated for quality control.

Approximately all of the differences were less than 5 percent, with the majority of them being less than 2 percent. CO_2 and NOx rates provide the most accurate estimates for all modes while other pollutants show mixed results. These results indicate that the resultant modal emission rates are valid and representative of observed emissions readings from PEMS equipment.

MOVES emissions rates for the corresponding model year of each individual vehicle are also included in Figure 14 through Figure 16. Please note that the definition of vehicle types used in MOVES encompasses a broad range of engine sizes and other characteristics that influence the amount of emissions. Therefore, MOVES emission rates should be treated as a representative of the average values for all different vehicles of a specific type and therefore discrepancy between observed emissions and fuel consumption rates and MOVES results are to be expected. The next section provides a discussion on comparing the observed emissions rates with MOVES estimates.

IMPACT ANALYSIS AND COMPARISON WITH MOVES

One of the main purposes of this study is to compare the observed emission rates with those from the EPA MOVES model. For this purpose, the MOVES model's modal emissions rates (emission rates for each operating bin) for each individual vehicle as well as each vehicle class were extracted. The MOVES and observed emissions rates were applied to two batches of drive cycles that were collected using GPS. These emissions estimates were then used to investigate the difference between the total amount of observed emissions and MOVES estimates.

The first batch of drive cycles used in this analysis consists of the actual speed profile of the vehicles during testing collected by a GPS unit. This first batch is referred to as *test cycles* in this report. The second batch of drive cycles represents the normal operation of their respective vehicle types, referred to as *operation cycles*. To collect speed data for these drive cycles, GPS units were installed on 18 of the City of Houston's fleet vehicles (eight Class 8 vehicles, five Class 6 vehicles, and five Class 4 vehicles) in Houston, TX, and second-by-second speed data were collected for approximately one week of their normal.

Valid speed data were extracted for each vehicle, and each vehicle's speed data were considered as one set (e.g., set 1, 2). Filtering criteria such as acceleration limits, maximum idle time before and after each moving section, and maximum and minimum speed thresholds were used in the process of extracting speed data. Figure 17 shows a sample test cycle and section of an operation cycle. Table 10 through Table 12 provide details of the operating cycles for all tested vehicle types.

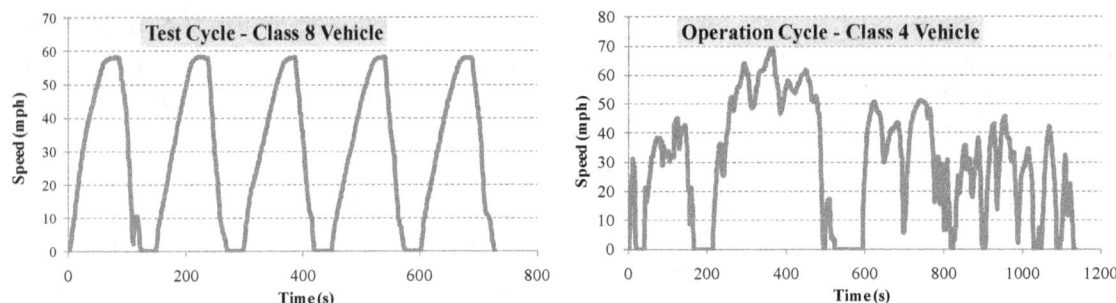

Figure 17. Example Drive Cycles: Test Cycle (Left), Operation Cycle (Right).

Table 10. Characteristics of Operating Cycles – Class 8 Vehicles.

	Drive Cycle Group							
	1	2	3	4	5	6	7	8
Vehicle Number	39621	39620	38418	38147	38166	37991	36919	30693
Total Duration (s)	9741	30673	18371	25373	17717	23883	13307	12269
Total Distance (mi)	53.3	169.1	108.6	127.2	128.6	104.2	68.6	59.4
Avg. Speed (mph)	19.7	19.8	21.3	18.0	26.1	15.7	18.6	17.4
Max Speed (mph)	60.3	57.3	65.9	66.6	63.7	69.1	66.6	63.9

Table 11. Characteristics of Operating Cycles – Class 6 Vehicles.

	Drive Cycle Sets				
	1	2	3	4	5
Vehicle Number	23697	25116	25225	37913	40073
Total Duration (s)	28632	21510	28573	5393	15781
Total Distance (mi)	127.9	115.5	186.3	21.1	81.5
Avg. Speed (mph)	16.1	19.3	23.5	14.1	18.6
Max Speed (mph)	59.1	58.8	61.2	64.3	68.7

Table 12. Characteristics of Operating Cycles – Class 4 Vehicles.

	Drive Cycle Sets				
	1	2	3	4	5
Vehicle Number	29408	29413	29417	40044	40087
Total Duration (s)	19487	15087	20980	3567	20739
Total Distance (mi)	127.1	70.3	150.9	22.2	86.7
Avg. Speed (mph)	23.5	16.8	25.9	22.4	15.1
Max Speed (mph)	70.6	60.1	72.6	60.5	59.8

The VSP value corresponding to each instance of the drive cycles was calculated based on Equation 1 utilizing MOVES default parameter values of Table 9 and a distribution of time spent in each operating mode bin was developed. The MOVES and observed emission rates for each individual vehicle were then applied to these operating mode distributions and the total observed and MOVES-estimated emissions for each vehicle were calculated. The average total emissions for each vehicle type were also calculated for a simple emission reduction benefit analysis. These estimates were then used to calculate the average emissions reduction benefits of replacing older vehicles with newer ones. The results of this analysis are discussed in the next section.

Figure 18 and Figure 19 show an example of drive cycle analysis for vehicle 8H-2 (or 21832). Figure 18 contains the results from test cycles whereas Figure 19 shows the total estimated emissions from operating cycles. Table 13 through Table 16 show the average distance-based emission rates in grams per mile (g/mi) for each operation cycle set, i.e., the average g/mi emissions from approximately one week of operation. The missing values in these tables are invalid or unavailable observations.

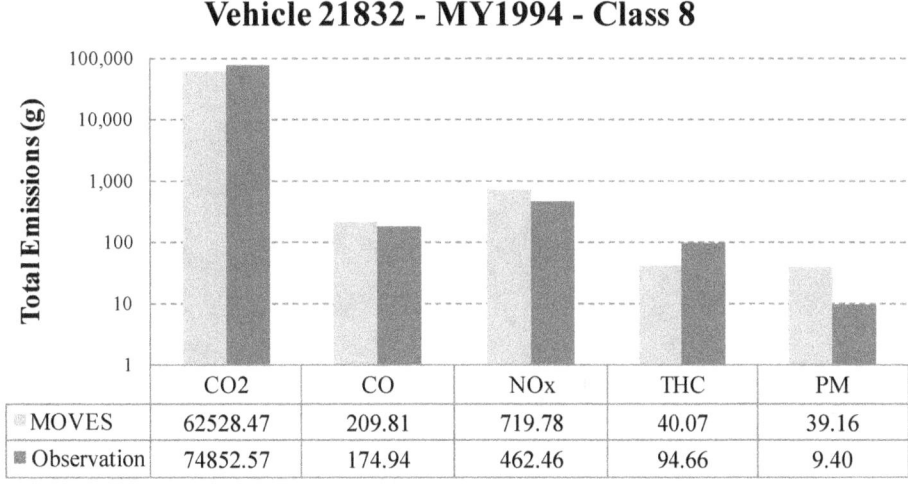

Figure 18. Total Emission for the Test Cycle of Vehicle 8H-2 (21832).

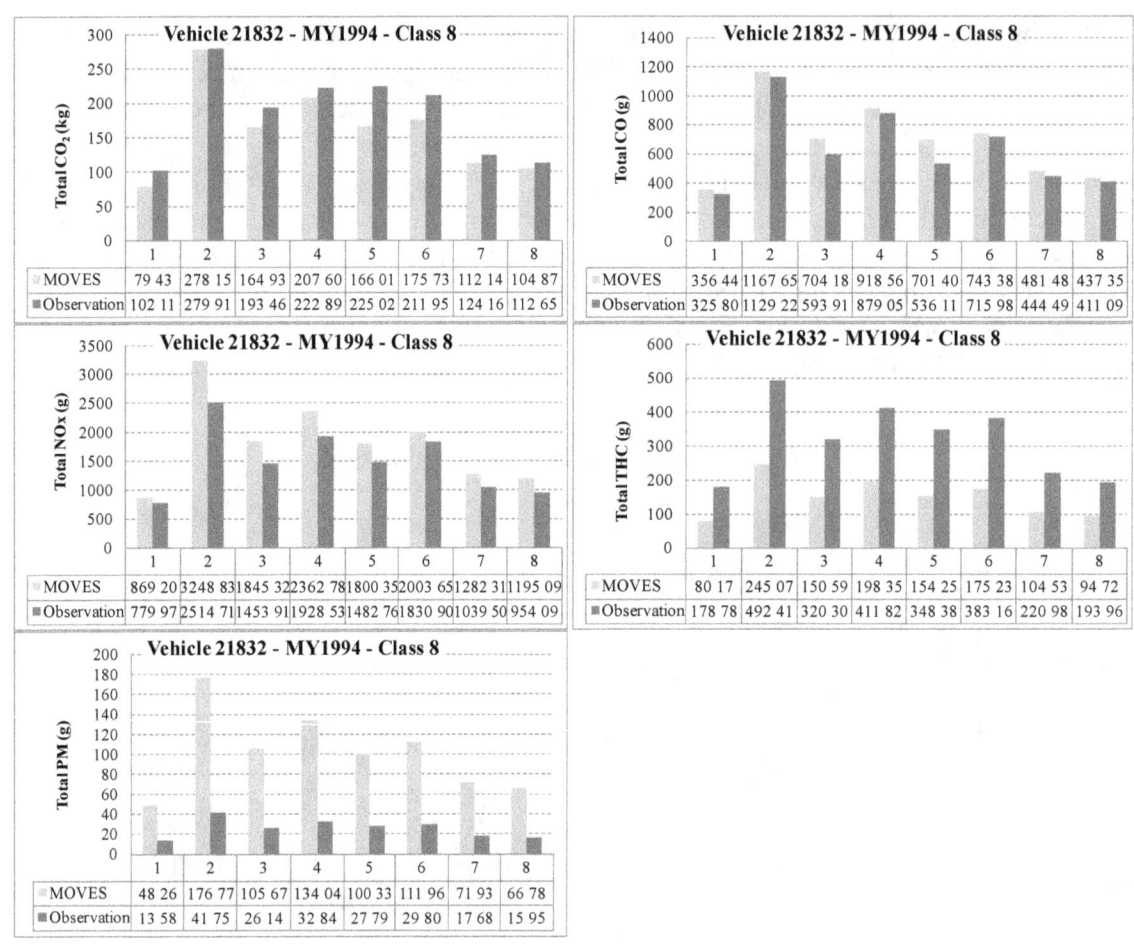

Figure 19. Total Emission for Operation Cycles of 8H-2 (21832).

Table 13. Observed Operation Cycles Emission Rates; High Emitting Class 8 Vehicles.

Vehicle #	Model Year	# of Cycle Sets	CO_2 (g/mi)	CO (g/mi)	NO_x (g/mi)	THC (g/mi)	PM (g/mi)
8H-1	1993	8	1668	4.56	15.3	8.00	
8H-2	1994	8	1797	6.15	14.6	3.11	0.25
8H-3	1995	8	1785	4.81	25.0	1.57	0.22
8H-4	1996	8	2049	4.71	20.8	6.55	
8H-5	2003	8	1375	6.07	10.6	0.82	0.08
		Average (All Vehicles)	1735	5.26	17.3	3.99	0.18
		Average (MY 2000 or Earlier Vehicles)	1825	5.06	18.9	4.78	0.23

Table 14. Observed Operation Cycles Emission Rates; Randomly Selected Class 8 Vehicles.

Vehicle #	Model Year	# of Cycle Sets	CO_2 (g/mi)	CO (g/mi)	NO_x (g/mi)	THC (g/mi)	PM (g/mi)
8-1	2003	8	1471	2.87	10.2	1.12	
8-2	2003	8	1706	4.62	12.4	0.90	0.19
8-3	2003	8	1650	6.60	12.3	0.76	0.12
8-4	2003	8	1700	5.41	13.4		0.22
8-5	2003	8	1529	3.89	12.5		0.22
8-6	2003	8	1810	4.26	14.9	0.66	0.12
8-7	2003	8	1700	3.58	13.3	0.44	0.12
8-8	2003	8	1722	9.13	12.5	0.41	0.23
8-9	2003	8	1781	12.26	14.6	0.34	0.21
8-10	2006	8			9.8		
8-11	2006	8	1614	5.17	9.0	0.87	0.13
8-12	2006	8	1754	6.16	10.3	0.98	
		Average (MY2003 Vehicles)	1674	5.85	12.9	0.66	0.18
		Average (MY2006 Vehicles)	1684	5.66	9.7	0.92	0.13
		Average (All Vehicles)	1676	5.81	12.1	0.72	0.17

Table 15. Observed Operation Cycles Emission Rates; Class 6 Vehicles.

Vehicle #	Model Year	# of Cycle Sets	CO_2 (g/mi)	CO (g/mi)	NO_x (g/mi)	THC (g/mi)	PM (g/mi)
6-1	1994	5	277	0.71	2.3		0.10
6-2	1999	5	1096	2.59	6.8	1.13	0.10
6-3	2000	5	1244	3.99	7.3	1.27	0.12
6-4	2000	5	1148	2.67	6.6	1.18	0.05
6-5	2001	5	475	1.28	2.2	0.14	0.06
6-6	2002	5	1374	5.30	10.9	0.39	0.09
		Average	936	2.76	6.0	0.82	0.09

Table 16. Observed Operation Cycles Emission Factors; High Emitting Class 4 Vehicles.

Vehicle #	Model Year	# of Cycle Sets	CO_2 (g/mi)	CO (g/mi)	NO_x (g/mi)	THC (g/mi)	PM (g/mi)
4H-1	1994	5	1535	2.23	5.5	0.52	
4H-2	1999	5	1951	3.72	7.7		0.13
4H-3	1999	5	939	1.71	4.9		0.01
4H-4	1999	5	1779	3.71	6.9	0.28	0.17
4H-5	1999	5	903		4.8	0.24	0.23
4H-6	1999	5	1184	2.32	5.9	0.11	
		Average	1382	2.74	6.0	0.29	0.14

Since test vehicles are of different makes and models, the researchers decided to use MOVES estimated emissions for the corresponding model year as the baseline for each individual vehicle. Using MOVES as a baseline makes it possible to compare results across different model years. Figure 20 through Figure 23 show the results of the cycle analysis for all vehicle types in terms of change from baseline MOVES estimates for all the operating cycles. Only operating cycle results are presented here since they represent emissions from the actual normal operation of the test vehicles. The average value for each pollutant is marked by a thick black horizontal bar. Note that Class 8 vehicles are divided into two categories: potential HEs based on opacity testing and randomly selected trucks with no opacity readings. These results are also presented in tabular format in the Appendix A.

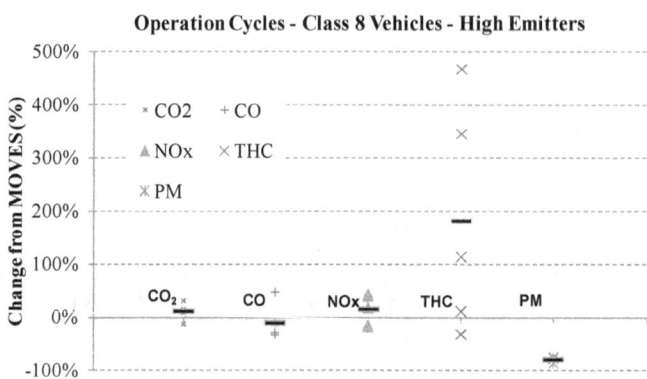

Figure 20. Cycle Analysis Results for High Emitting Class 8 Vehicles.

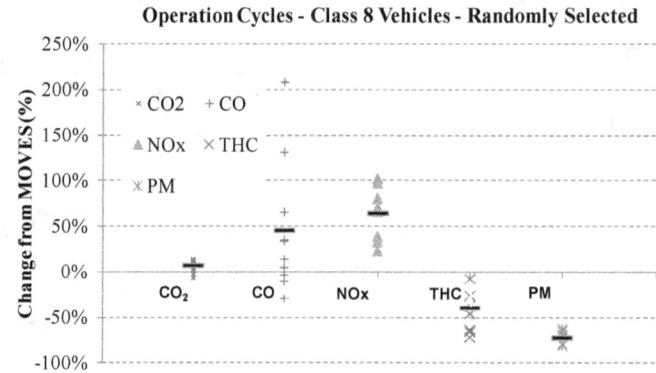

Figure 21. Cycle Analysis Results for Randomly Selected Class 8 Vehicles.

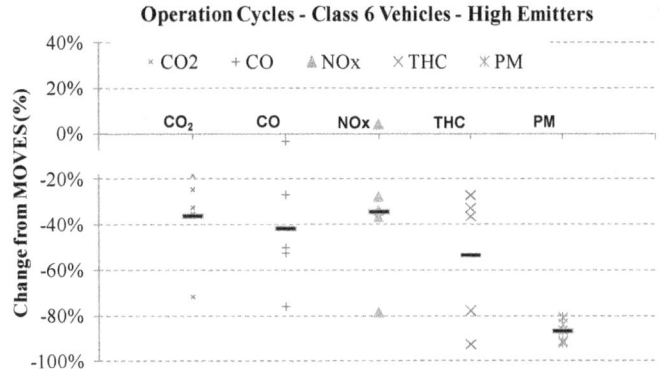

Figure 22. Cycle Analysis Results for Class 6 Vehicles.

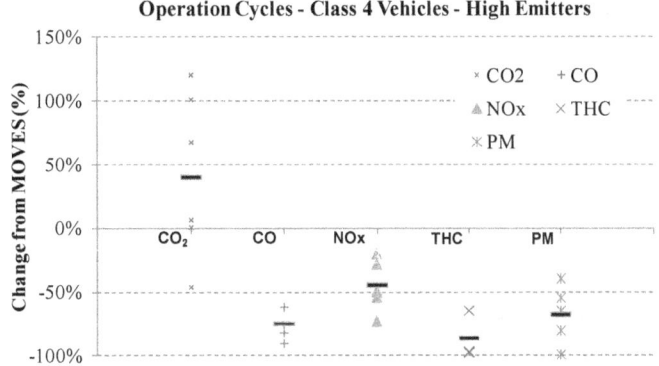

Figure 23. Cycle Analysis Results for High Emitting Class 4 Vehicles.

Overall Results

In general, observed CO_2 rates have the highest consistency with MOVES rates. This is not surprising given the nature of CO_2 formation in the engine and its relative insensitivity between model years. The researchers observed that CO_2 emissions of Class 8 vehicles, both categories of potential high emitting vehicles and randomly selected ones, have the lowest divergence from MOVES rates. This lowest divergence indicates low variability for this class of vehicles in terms of fuel consumption.

The observed emission rates of Class 6 vehicles were always lower than rates predicted by MOVES indicating that the tested vehicles were probably of the lower boundary of fuel consumption for this vehicle class. As it is shown in Figure 23, Class 4 vehicles on the other hand showed higher CO_2 rates than MOVES estimates, which can be interpreted that they belong to the higher boundary of this vehicle type when it comes to fuel consumption characteristics. This is specifically evident for high VSP operation mode bins, i.e., high load bins.

CO and NOx emissions from potential high emitting vehicles, which are generally older vehicles, showed high consistency with MOVES estimates. Randomly selected vehicles, which are of newer model years (2003 and 2006), had a higher variability in terms of divergence from MOVES estimates. This variability might be the result of the limited field data used in MOVES for HDDVs; only one MY 2007 vehicle was included in the sample vehicle class while the rest of the sample being older than model year 2001 (*24*). With only one exception, THC emissions from potential high emitting Class 8 vehicles were higher than MOVES estimates; however, THC emissions from all the other vehicle type were lower than MOVES. The observed PM emissions for all test vehicles were significantly lower than MOVES estimates, potentially because of the difference in measurement methodologies used for MOVES (filter samples during dynamometer testing) and in this study (light scattering detection during on-road PEMS testing).

IDLING TESTING RESULTS

The research team conducted all idling testing inside the EERF test chamber under the same conditions. The temperature for each test was set to 86°F, and the RH was set to 60 percent. Each vehicle idled for approximately one hour at the conditions, using the default (low) idling engine speed of the test vehicle.

Class 8 HEs

A total of 6 Class 8 HEs, as described in Table 5, were tested. Figure 24 through Figure 26 show the gaseous, PM, and aldehyde emissions rates, along with fuel consumption, from each of the Class 8 HEs.

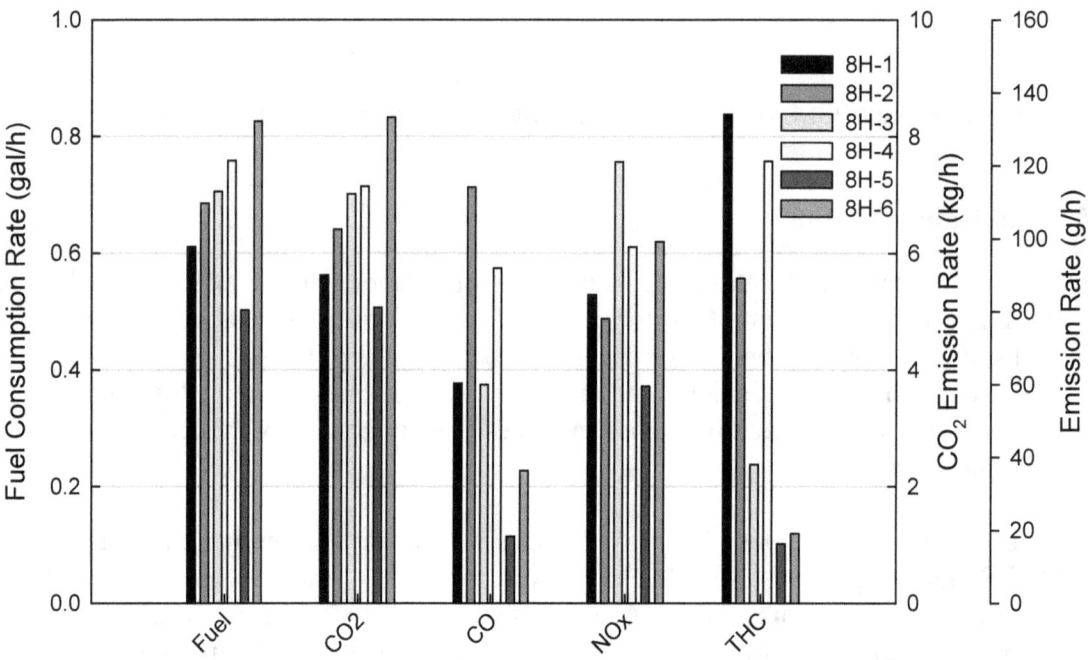

Figure 24. Class 8 HEs Idle Emissions.

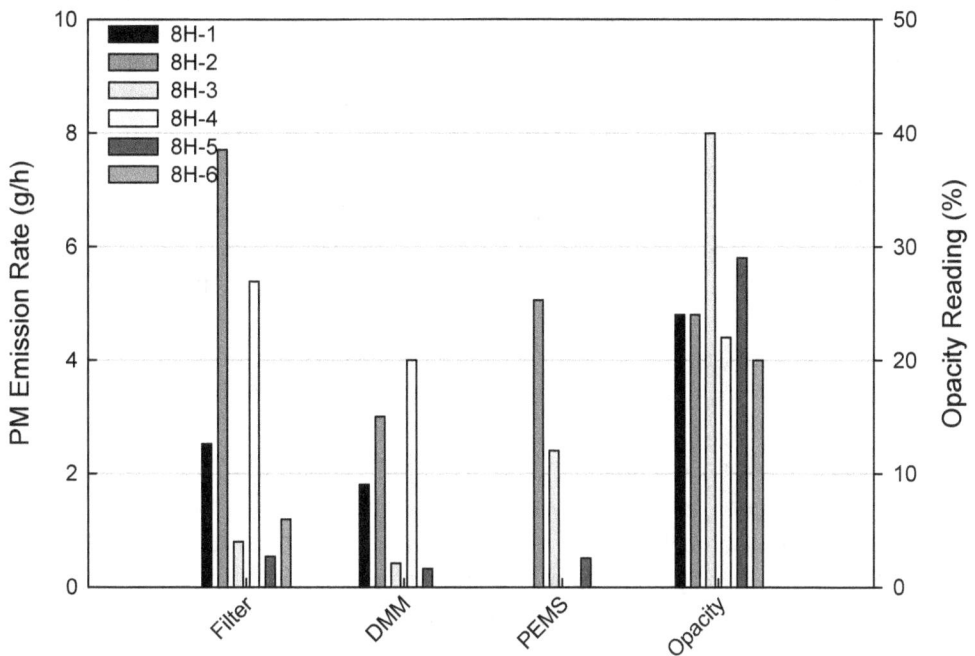

Figure 25. Class 8 HEs PM Emissions.

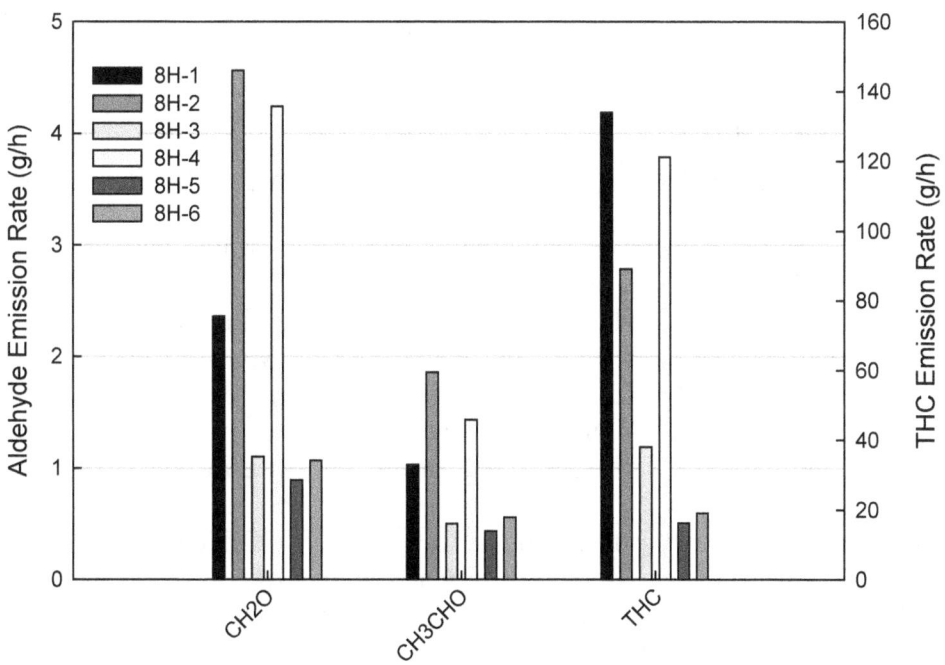

Figure 26. Class 8 HEs Aldehyde Emissions.

As the graphs show, for both fuel consumption and CO_2 and other pollutant emissions, there is a wide range of different values among different vehicles, which indicates that different vehicles show different combustion characteristics. When compared to the expected emissions values from MOVES, as shown as percentages (observed/expected) in Table 17, each vehicle showed emissions lower than estimated for CO_2 and with one exception showed lower PM rates. NOx comparisons results were split with two vehicles being higher than the estimated, three lower, and one equal to the MOVES estimate. CO, THC, and aldehyde emission rates were higher than MOVES estimates, with the exception of one vehicle each for CO (8H-5) and for aldehydes (8H-3), respectively, although the magnitudes of the differences were large—for example, for THC, the highest percentage was 801 percent while the lowest was 154 percent.

One important note when looking at the MOVES comparison is that the percentages do not represent actual emissions measured, they are simply a percentage of the expected value from MOVES. This is due to the fact that the vehicles are from different model years and therefore have different MOVES estimates. Because of this the percentage numbers presented in Table 17 should not be used to compare one vehicle to another for actual emission rates. For example, in Table 17 the NOx percentage for vehicle 8H-6 (211 percent) is much higher than that for vehicle 8H-3 (100 percent). However, by looking at the emissions rates in Figure 24, we see that vehicle 8H-3 had a higher NOx emission rate than that of vehicle 8H-6.

Table 17. Class 8 HEs Percentage of MOVES Emissions.

Vehicle #	CO_2	NOx	CO	PM(filter)	PM (DMM)	PM (PEMS)	THC	CH_2O	CH_3CHO
8H-1	60%	70%	164%	58%	42%	NA	801%	169%	201%
8H-2	69%	65%	311%	113%	44%	74%	517%	318%	351%
8H-3	75%	100%	163%	12%	6%	35%	225%	78%	97%
8H-4	76%	81%	251%	79%	59%	NA	699%	294%	269%
8H-5	53%	126%	54%	9%	6%	9%	154%	101%	134%
8H-6	87%	211%	108%	20%	NA	NA	181%	121%	173%

Figure 25 and Figure 26 show the Class 8 HEs' emission results for PM and aldehyde, respectively. During the idle testing, the researchers measured PM with three different methodologies. The "filter" results show the measurements from filter samples taken and weighed based upon the current federal test procedure. The DMM results were measured by DMM. Particles entering DMM were charged. The charged particles were collected on faraday cups based upon their sizes. The particle charges were then converted into a PM mass. The PEMS results were the converted PM mass of the particles detected by a laser light scattering.

All PM measurements from three different methodologies were conducted simultaneously with all three different PM measurement systems. Compared to the MOVES PM estimates, with the exception of the filter measurement result of 8H-2, all PM emission rates were lower than the MOVES estimates as shown in Table 17.

Figure 26 shows the aldehyde emission results from the Class 8 HEs, along with the THC emissions. Since the aldehydes are parts of the THC emissions, THC emissions are included in the graph for comparison purposes. As shown in Figure 26, aldehyde emission rates were only a small percentage of THC emission rates. When comparing these results to MOVES in Table 17 we see that the results observed in the study are much higher than the estimated rates for aldehydes, with the exception of vehicle 8H-3.

Class 8 Randomly Selected Vehicles

A total of 12 Class 8 vehicles were randomly selected for testing, as outlined in Table 6. These vehicles were selected from either MY 2003 or MY 2006. Figure 27 through Figure 29 show the gaseous, PM, and aldehyde emissions rates from each of the Class 8 randomly selected vehicles. Due to a malfunction of one of the PEMS some of these vehicles do not have THC results.

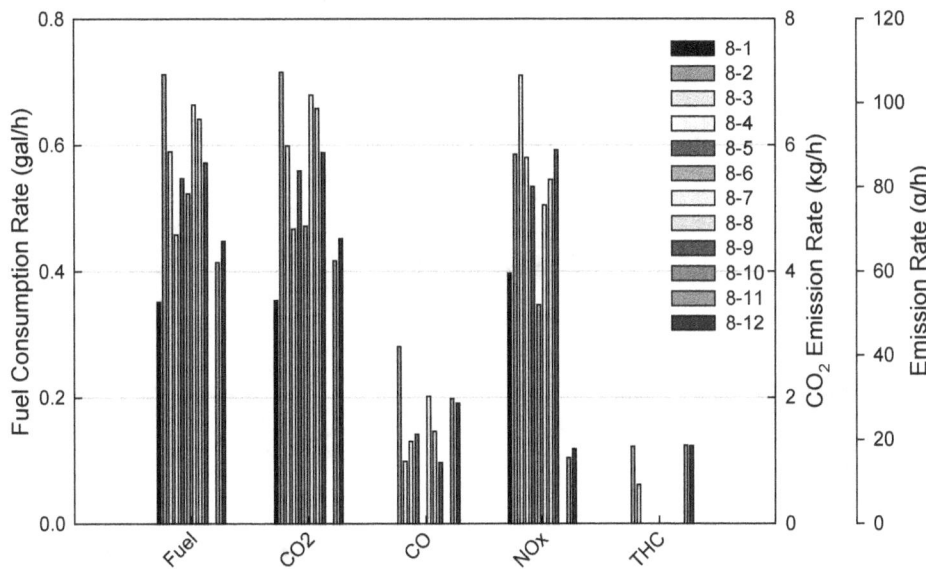

Figure 27. Class 8 Randomly Selected Vehicles Idle Emissions.

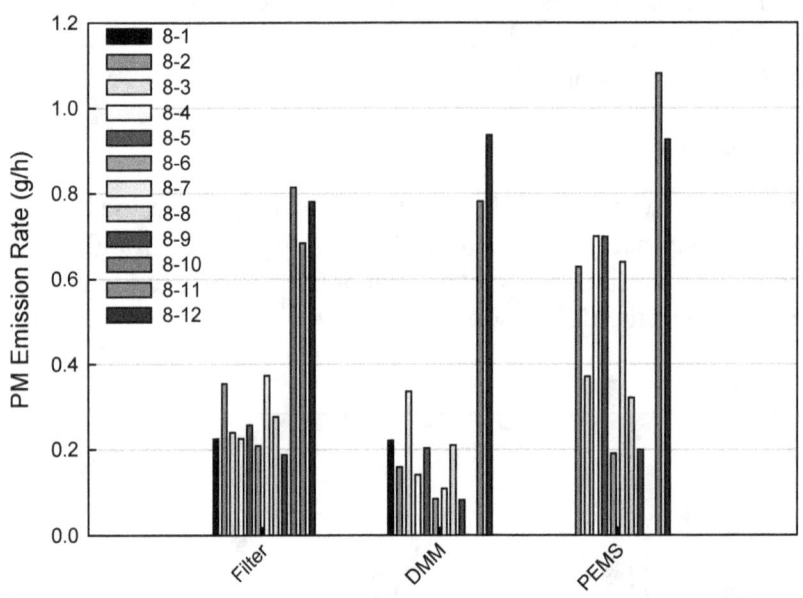

Figure 28. Class 8 Randomly Selected PM Idle Emissions.

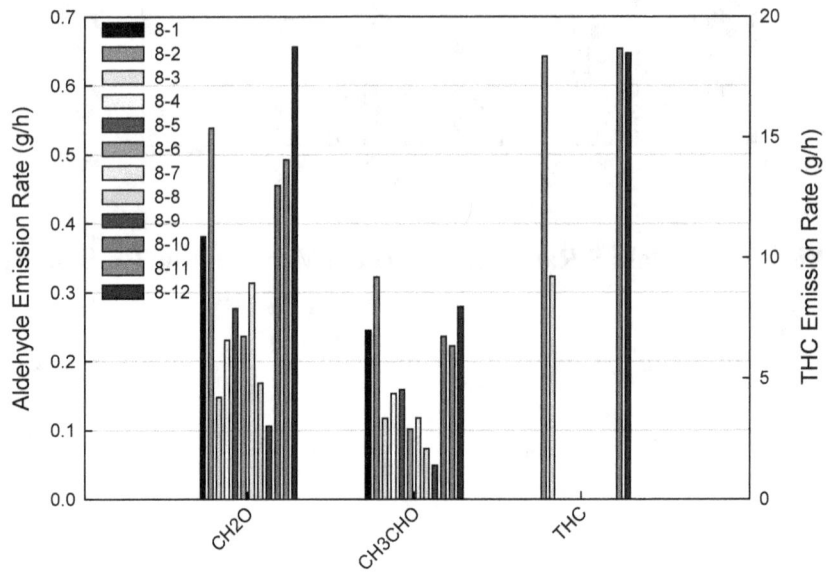

Figure 29. Class 8 Randomly Selected Aldehyde Idle Emissions.

Figure 27 shows the results from the randomly selected Class 8 vehicles. Similar to the results from the HEs, they have a wide range of different values among different vehicles. However, when compared to the MOVES estimation rates (Table 18), we see that with the exception on NOx, emission rates of CO_2, and all other pollutants from almost all of the randomly selected vehicles were under the estimated emissions rates. The results for NOx were different, with only two of the 12 tested vehicles being under the estimated MOVES rates.

Table 18. Class 8 Randomly Selected Vehicles Percentage of MOVES Emissions.

Vehicle #	CO_2	NOx	CO	PM(filter)	PM (DMM)	THC	CH_2O	CH_3CHO
8-1	37%	127%	NA	4%	3%	NA	43%	76%
8-2	75%	187%	125%	6%	6%	174%	61%	99%
8-3	63%	227%	44%	4%	3%	87%	17%	36%
8-4	49%	185%	58%	4%	2%	NA	26%	47%
8-5	59%	171%	63%	4%	3%	NA	31%	49%
8-6	49%	111%	NA	4%	1%	NA	27%	31%
8-7	71%	161%	90%	6%	2%	NA	36%	36%
8-8	69%	174%	65%	5%	4%	NA	19%	22%
8-9	62%	189%	43%	3%	1%	NA	12%	15%
8-10	NA	NA	NA	14%	NA	NA	53%	75%
8-11	44%	33%	91%	12%	13%	181%	57%	70%
8-12	47%	38%	88%	13%	16%	180%	77%	88%

When we compare the randomly selected vehicles to the Class 8 higher emitters we see that, in most instances, the HEs showed a higher percentage of MOVES estimation rates.

The PM and aldehyde measurements from the randomly selected Class 8 vehicles are shown in Figure 28 and Figure 29, respectively. When the observed emission rates are compared to MOVES rates, the observed aldehyde emission rates of all 12 vehicles were lower than those of the estimated (MOVES) rates. For THC, three of the four vehicles (the others were not reported due to instrument malfunction) observed had emission rates higher than the estimated rates. This result is much different from the results of the HEs. Compared to THC emission rates, aldehyde emission rates of randomly-selected vehicles are also very low, similar to the high-emitters discussed in the previous section.

Class 6 Vehicles

The researchers selected six Class 6 vehicles testing, as outlined in Table 7. Originally these vehicles were to be labeled as HEs. During the opacity testing process, however, it was found that none of the Class 6 vehicles tested qualified as a HE (opacity readings were under the criterion of 20 percent). Once this was discovered, the research team determined that the Class 6 vehicles with the highest opacity reading would be tested, but they would not be labeled HEs.

Figure 30 below shows the emissions rates for the Class 6 vehicles. As seen with the other classes, the emissions of the vehicles were varied when looking at the emission rates. When we factor in the comparison to MOVES, as shown in Table 19, we see that all of the observed emission rates are under the MOVES estimated rates, with only one exception of NOx for a vehicle (6-6).

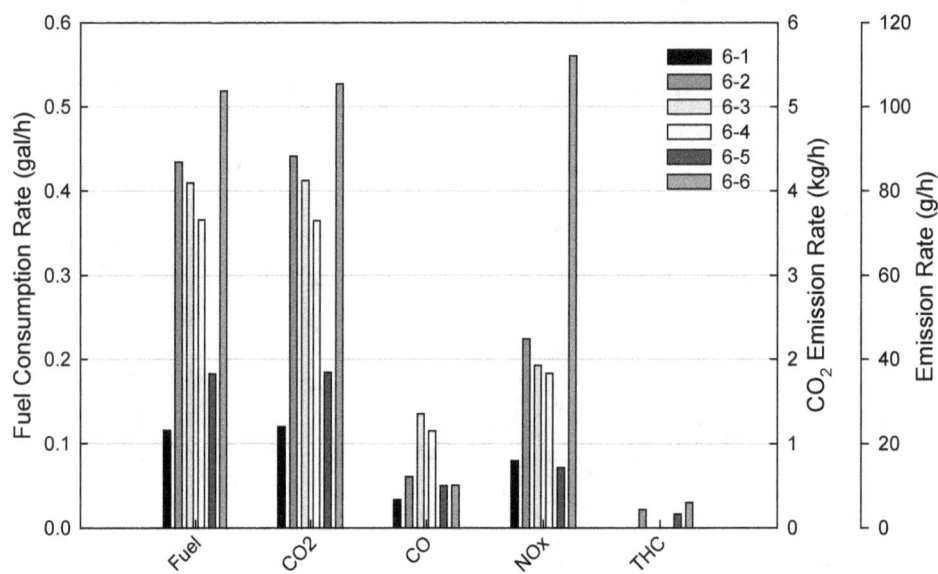

Figure 30. Class 6 Idle Emissions.

Table 19. Class 6 Percentage of MOVES Emissions.

Vehicle #	CO$_2$	NOx	CO	PM(filter)	PM (DMM)	THC	CH$_2$O	CH$_3$CHO
6-1	13%	13%	19%	11%	5%	NA	15%	21%
6-2	46%	47%	43%	9%	17%	25%	17%	23%
6-3	43%	40%	97%	9%	10%	NA	29%	36%
6-4	38%	38%	82%	6%	9%	NA	34%	40%
6-5	19%	15%	36%	2%	5%	18%	14%	17%
6-6	55%	116%	36%	3%	5%	34%	17%	21%

Figure 31 shows the PM emissions along with opacity readings for the selected Class 6 vehicles. Figure 32 shows the aldehyde and THC emission rates for the selected vehicles. Similar to the other classes mentioned in earlier sections, aldehyde emission rates are very low as compared to those of THC.

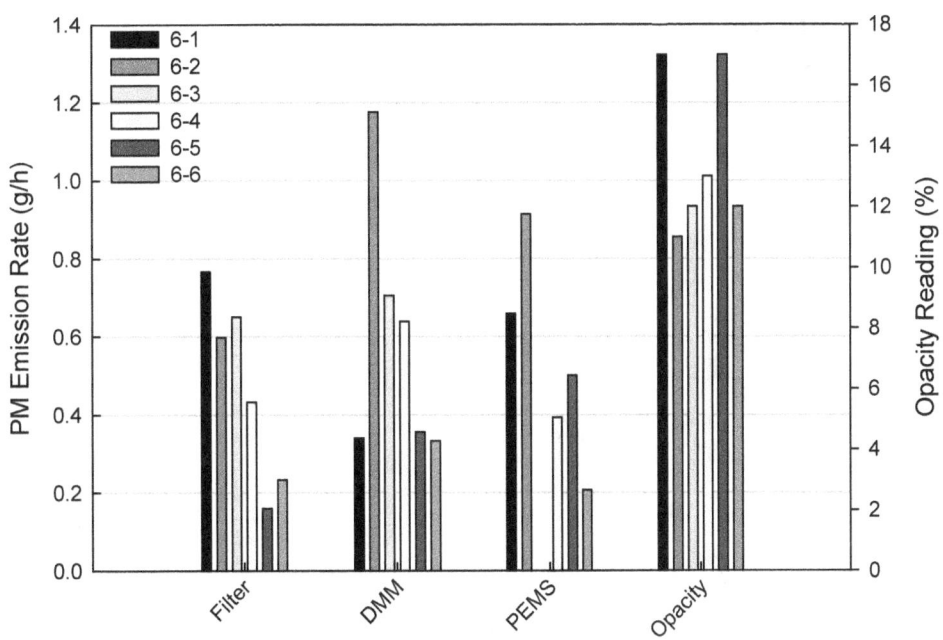

Figure 31. Class 6 PM Idle Emissions.

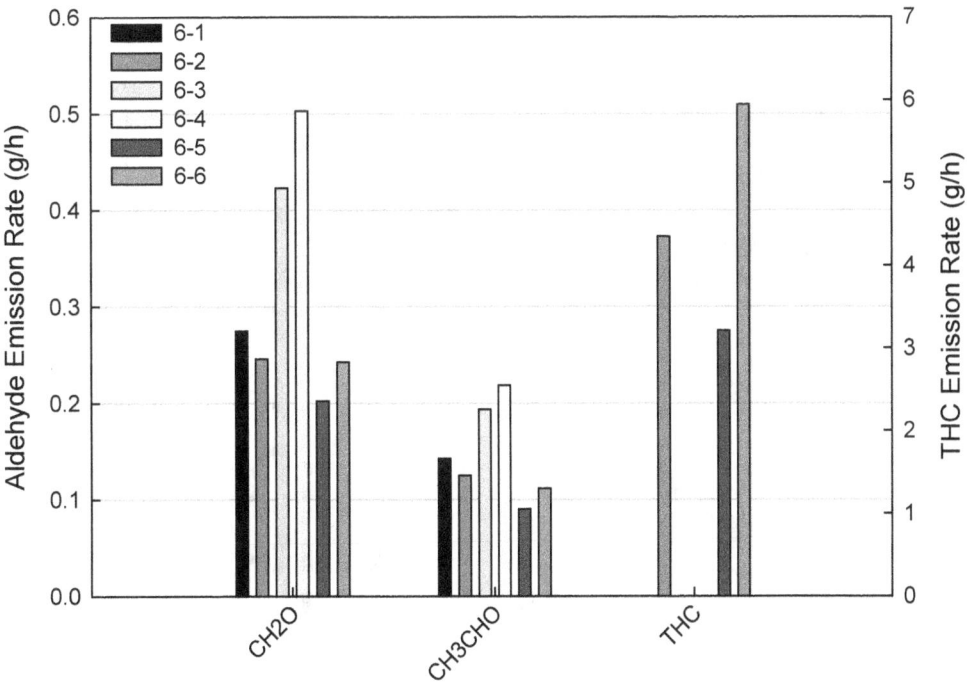

Figure 32. Class 6 Aldehyde Idle Emissions.

Class 4 HEs

Six Class 4 HEs, as described in Table 8, were tested. Figure 33 through Figure 35 show the gaseous, PM, and aldehyde emissions rates from each of the Class 4 HEs.

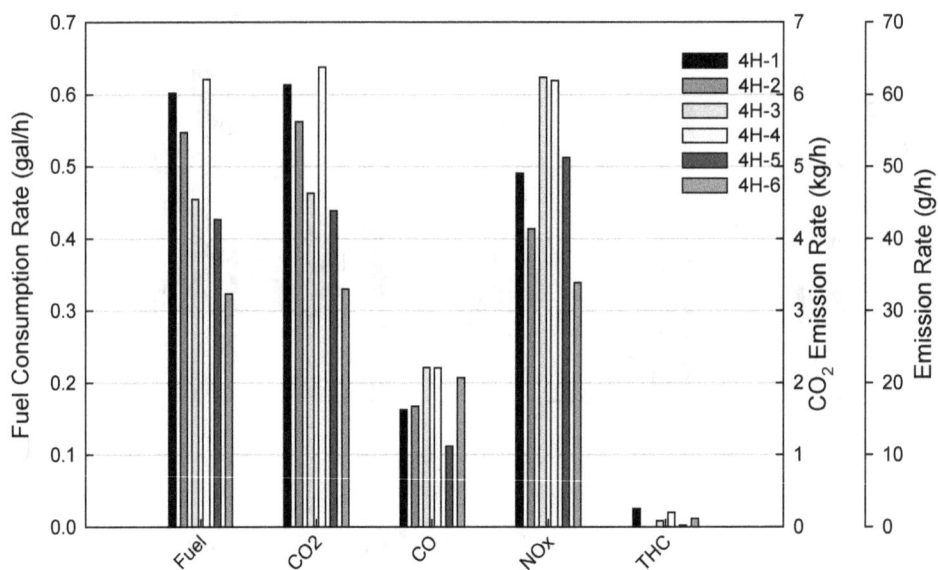

Figure 33. Class 4 HEs Idle Emissions.

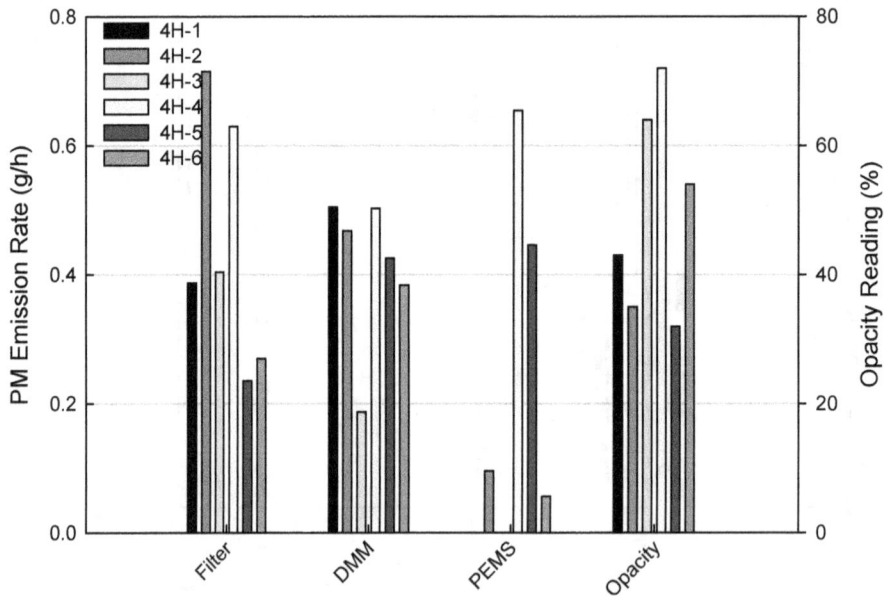

Figure 34. Class 4 HEs PM Idle Emissions.

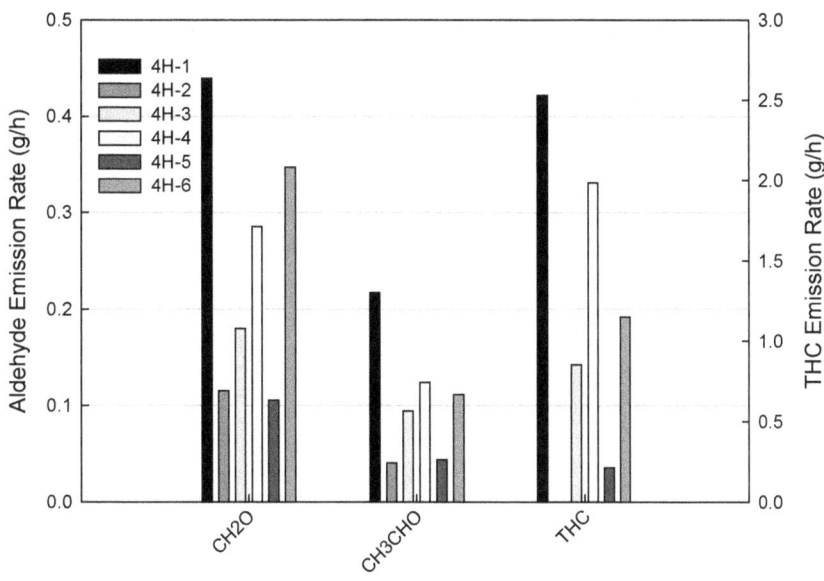

Figure 35. Class 4 HEs Aldehyde Idle Emissions.

Wide ranges of variation were observed for the measure values among vehicles in this class, as is similar to other classes mentioned in earlier sections. As shown below in Table 20 the comparison to MOVES shows that most of the observed emission rates of the class 4 HEs were lower than the MOVES estimates. The one exception was observed for CO_2, where the observed emission rates were higher for half the vehicles as shown in Table 20. The other three vehicles were at or near the estimated rates from MOVES. Note that for Class 4 HEs, aldehyde emission rates take significant parts of THC emission rates as shown in Figure 35. For example, for vehicles 4H-4, CH_2O emission rate (0.105 grams per hour [g/h]) is 50 percent of THC emission rate (0.21 g/h), and CH_3CHO rate (0.044 g/h) is about 20 percent of the THC.

Table 20. Class 4 HEs Percentage of MOVES Emissions.

Vehicle #	CO_2	NOx	CO	PM(filter)	PM (DMM)	PM (PEMS)	THC	CH_2O	CH_3CHO
4H-1	142%	15%	20%	5%	7%	NA	14%	29%	39%
4H-2	119%	15%	22%	11%	7%	2%	NA	8%	8%
4H-3	98%	22%	29%	6%	3%	NA	5%	13%	18%
4H-4	135%	22%	29%	10%	8%	10%	12%	20%	24%
4H-5	93%	18%	15%	4%	7%	7%	1%	7%	8%
4H-6	70%	12%	27%	4%	6%	1%	7%	25%	21%

Overall Results

Figure 36 through Figure 38 show the average emission rates of each class of vehicles tested along with the associated confidence intervals as bars. As Figure 36 through Figure 38 show the average rates for Class 8H are higher than the regular Class 8 vehicles as well as all other classes in every category. For PM, Class 6 HDDVs showed the next highest rates followed by Class 4 HEs, and Class 8 randomly selected HDDVs with the least. For THC and aldehydes, Class 8 was the next highest, followed by Class 6. Class 4HEs' emission rates were lowest. For all other emissions as well as fuel consumption, Class 8 was the next highest, followed by Class 4H and then Class 6.

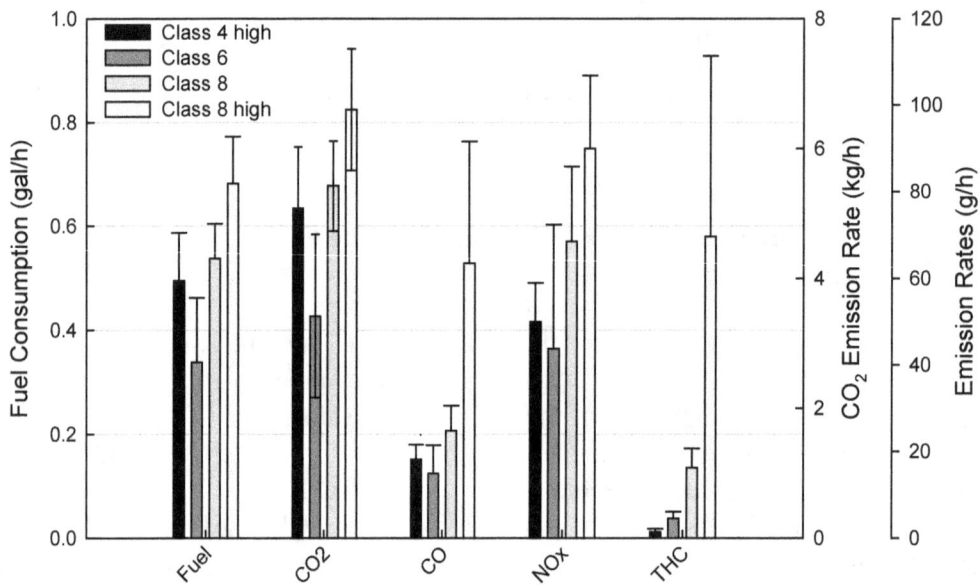

Figure 36. Comparison of Emissions Rates of Vehicle Classes.

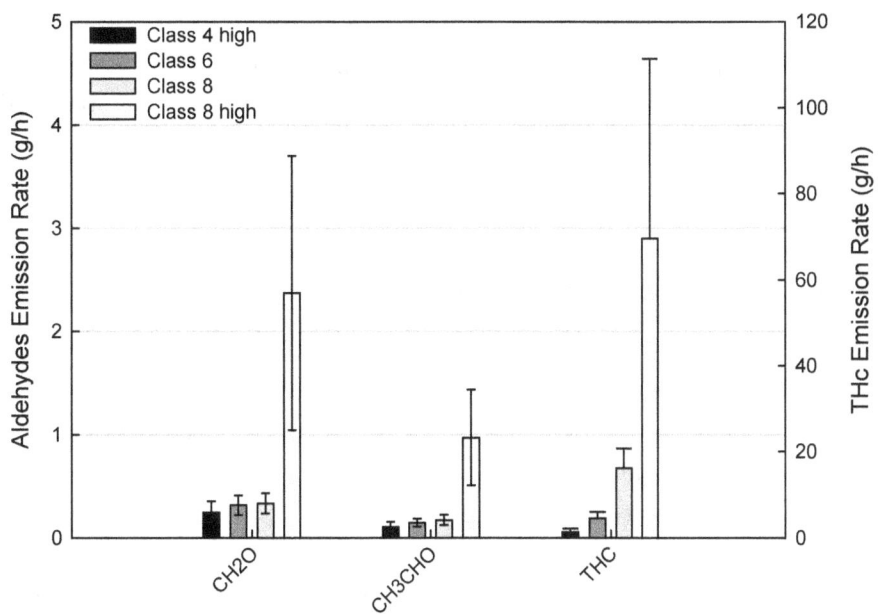

Figure 37. Comparison of Aldehyde Emissions Rates of Vehicle Classes.

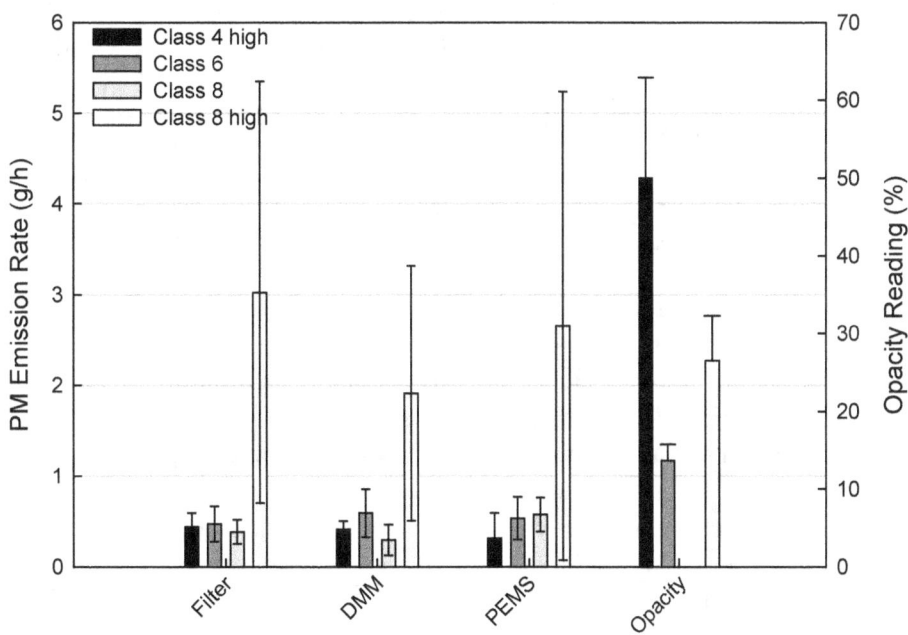

Figure 38. Comparison of PM Emissions Rates of Vehicle Classes.

EMISSIONS REDUCTION BENEFIT ANALYSIS

The research team developed a simple methodology to estimate average emission reduction benefits of utilizing emissions reduction strategies such as repowering, retrofit, and replacement on potential high emitting vehicles. The methodology is based on the data collected for this study as well as MOVES emissions rates.

The main assumptions of the estimation methodology are as following:
- Only the average impact of the emissions reduction strategies (ERS) are considered; i.e., it is assumed that the overall improvement resulted from deploying an ERS on a high emitting vehicle will bring the average emission level of that vehicle to a newer model year emission rates.
- Two different improved conditions are selected to provide a low and a high range of potential emissions reduction benefits. MY 2006 is used to calculate low estimate and MY 2011 is used for the high estimate.
- Observed emission rates from the operation cycles (Table 13 through Table 16) are used in the calculations. If observations are not available for improved conditions (i.e., MY 2006 and 2011), MOVES emissions rates for those model years are used instead. Table 21 shows MOVES rates used in this analysis.

Table 21. MOVES Rates Used for Emissions Reduction Analysis - Analysis Year 2011.

	CO_2 (g/mi)	CO (g/mi)	NO_x (g/mi)	THC (g/mi)	PM (g/mi)
MY 2011 Class 8 Vehicle	1602	0.42	0.97	0.13	0.02
MY 2011 Class 6 Vehicle	1657	0.43	0.84	0.13	0.02
MY 2006 Class 6 Vehicle	1655	4.23	7.64	1.26	0.55
MY 2011 Class 4 Vehicle	1104	1.19	0.81	0.07	0.01
MY 2016 Class 4 Vehicle	1217	5.34	6.56	0.80	0.28

The high estimate results presented in Table 22 indicate that replacing an old vehicle (older than 2000) with a vehicle that adheres to 2011 emissions standards will significantly reduce almost all emission for all three vehicle classes. The only exception to this trend is CO_2 emissions for Class 6 vehicles of which the observed rates were significantly higher than MOVES estimates. The low estimates based on replacing the old vehicle with a vehicle complies with 2006 emissions standards, on the other hand, show modest reductions for Class 8 vehicles. The empty cells in the table indicate that a reduction could not be calculated, mainly because of lack of observations for the representative model year of 2006. Based on the data collected in this study, CO emissions for Class 8 vehicles show no reduction benefit for an upgrade to MY 2006 level.

Table 22. Potential Emissions Benefits by Replacing MY 2000 or Earlier Vehicles.

		CO_2	CO	NO_x	THC	PM
Class 8 Vehicle	Low Estimate	−2.9%		−43.8%	−76.9%	−27.8%
	High Estimate	−7.7%	−92%	−95.0%	−96.8%	−88.4%
Class 6 Vehicle	Low Estimate					
	High Estimate	−11.2%	−67.8%	−89.6%	−92.8%	−88.7%
Class 4 Vehicle	Low Estimate					
	High Estimate	−12.0%	−51.2%	−85.4%	−76.2%	−92.5%

In addition to looking at the reduction strategies another option to reduce emissions strategy is to remove the HEs from the fleet. The idling data, as shown in Table 23, show that, by removing the Class 8 HEs from the fleet and allowing the random vehicles to do the same work, emissions would be reduced anywhere from 18 to 87 percent in addition to a fuel savings of 21 percent.

Table 23. Potential Fuel Savings and Emissions Benefits by Replacing HEs with Normal Vehicles.

	Fuel Consumption Rate (gal/hr)/Emissions Rates (g/h)							
	Fuel	CO_2	NO_x	CO	PM (filter)	THC	CH_2O	CH_3CHO
Random Class 8 Vehicles	0.54	5418	68.44	24.79	0.39	16.18	0.33	0.17
Class 8 HEs	0.68	6595	89.95	63.46	3.03	69.61	2.37	0.77
Fuel Saving/Emission Reduction	21%	18%	24%	61%	87%	77%	86%	82%

CORRELATION OF DIFFERENT PM MEASUREMENTS

When analyzing the test results, the research team examined whether there was a correlation between the opacity readings and the emission rates, especially PM rates. Figure 39 below shows the measured opacity readings of different vehicles and PM emissions rates measured from the same vehicles with using PM filters during the idle testing. A quick review of the graph shows that there is no true relationship between the opacity readings obtained in the process of the HE selection and the PM emission rates of those vehicles measured during the idling testing. One potential reason for this is the fact that the opacity readings based on the SAE J1667 test procedures were taken during the idling at high idle speeds (more correctly open throttle idling) while PM samples were taken during the idling emissions testing at low speed idle. Further study with same idling conditions is required to verify any correlations between the opacity readings and PM measurements.

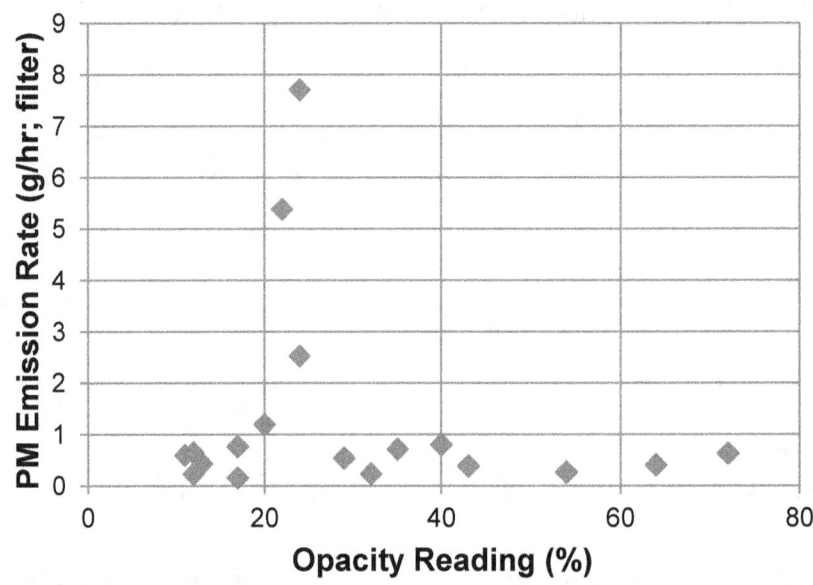

Figure 39. Correlation between PM Filter Emission Rates and Opacity Readings.

Also, different measurement methods use different methodologies to measure PM. For filter sampling, the current federal test procedure, the exhaust particles are collected on a filter. Then the collected PM mass is determined by weighing the filters before and after sampling. For DMM, electrical low pressure impactor (ELPI) method was used. For the ELPI method, exhaust particles entering into DMM are charged, and the charged particles are collected on faraday cups, charge measurement devices, depending on aerodynamic sizes of the particles. The measured charges are then converted into PM mass. For PEMS, the exhaust particles are detected by a laser light scattering. Finally, the scattered light measurements are converted to PM mass using PEMS internal software calculation.

Among these PM measurement methods, some correlations were found. As shown in Figure 40 and Figure 41, PM emission rates measured by DMM were correlated with those measured by filter samples for Class 8 vehicles: $R^2 = 0.764$ for Class 8 HEs and 0.858 for Class 8 randomly selected HDDVs. Also, similar correlations between those by PEMS and by filter samples were observed as shown in Figure 42 and Figure 43: $R^2 = 0.851$ for Class 8 HEs and 0.653 for Class 8 randomly selected HDDVs. More systematic studies with different test conditions would facilitate these findings.

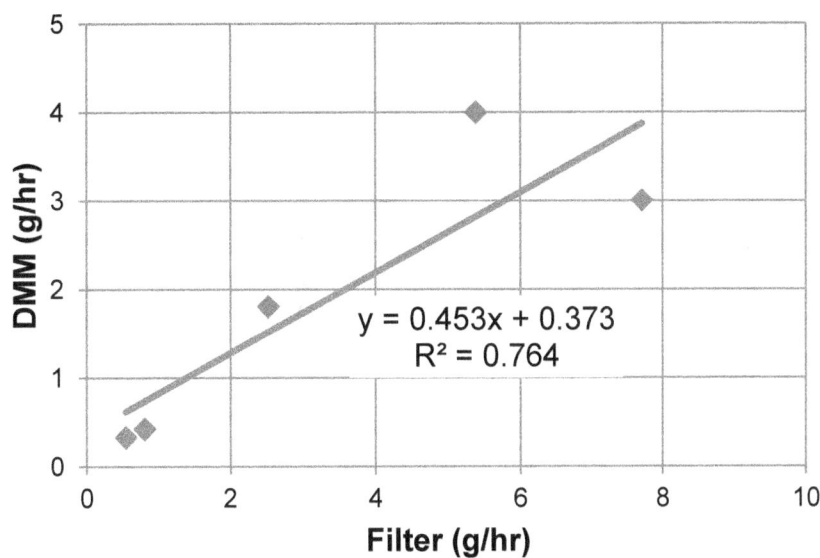

Figure 40. Correlation between Filter and DMM PM Emission Rates (Class 8H Vehicles).

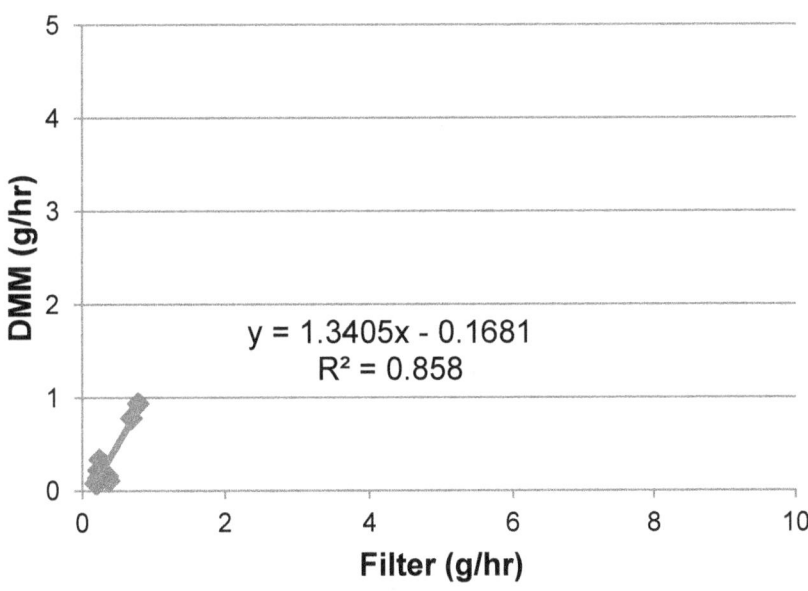

Figure 41. Correlation between Filter and DMM PM Rates (Class 8 Vehicles).

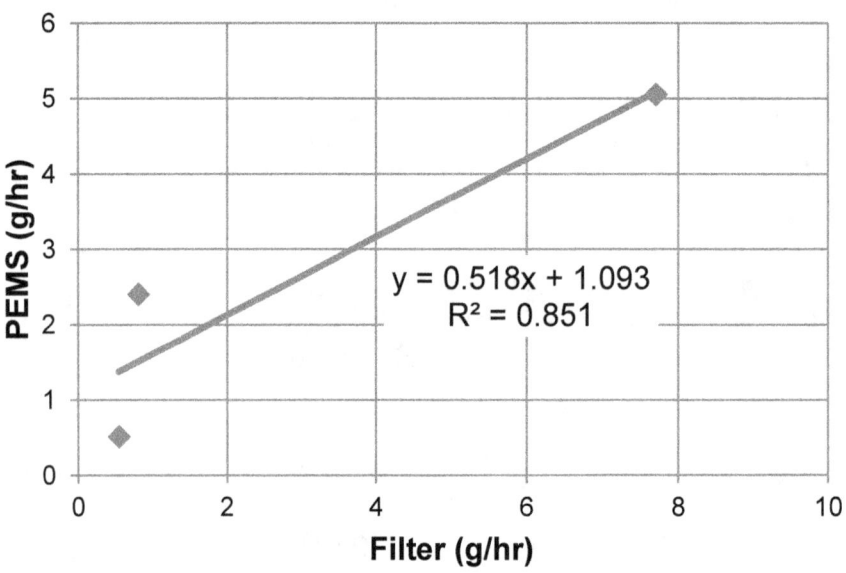

Figure 42. Correlation between Filter and PEMS PM Rates (Class 8H Vehicles).

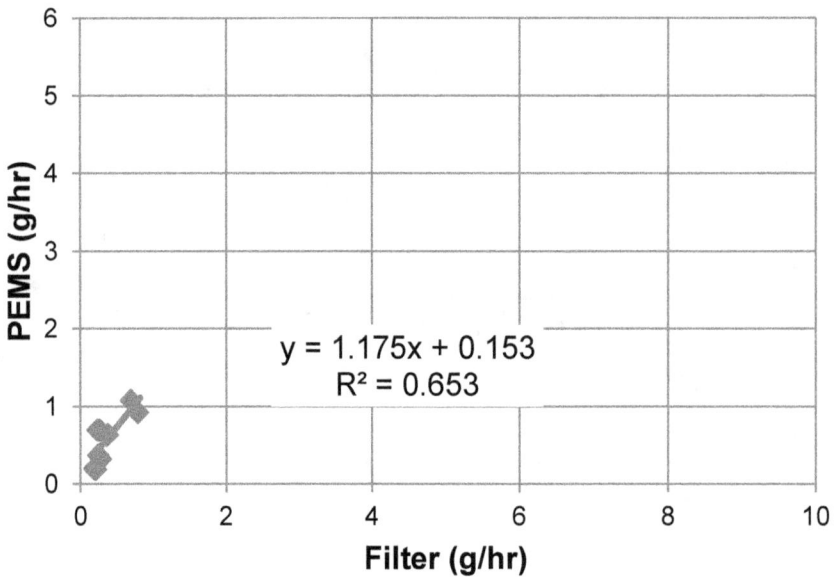

Figure 43. Correlation between Filter and PEMS PM Rates (Class 8 Vehicles).

CHAPTER 5: CONCLUSIONS

This project characterized the in-use emissions of HEs and vehicles representative of the general HDDV fleet and compared their emissions. Following a literature review and state of practice assessment, the research team developed an approach to identifying high emitter vehicles through a process of opacity testing. Vehicles from the City of Houston's fleet were screened using opacity testing, and 12 HEs were selected for emissions testing, along with 18 vehicles randomly selected to be representative of the overall HDDV fleet (i.e., non-HEs). The test vehicles included the following vehicle classes and types:

- 6 HEs in Class 8.
- 12 randomly selected MY 2003 and 2006 vehicles in Class 8 (representing non-high emitters/normal HDDVs).
- 6 HDDVs in Class 6 (representing non-high emitters/normal HDDVs).
- 6 HEs in Class 4.

Using the 30 selected HDDVs, driving and idling emissions tests were performed to characterize HEs and other normal HDDV emissions. The emissions results measured by PEMS were compared among vehicles and between vehicle classes, and HEs and normal HDDVs, respectively. In general, emissions of the HDDVs had a high variability between them even within a vehicle class.

Driving testing was performed by following the developed drive cycles, which was designed to show emissions with respect to driving characteristics such as vehicle speeds, accelerations, and engine loads. Consequently, driving testing analysis was focused more on comparisons between the observed emissions and MOVES-estimated emissions. The major findings of the driving testing are:

- MOVES estimates for CO_2, CO, and NOx are best representative of observation for potential Class 8 HEs, which tend to be older vehicles.
- THC emissions of the potential Class 8 HEs were generally higher than MOVES estimates indicating a potential relationship with opacity readings.
- Newer Class 8 vehicles (MYs 2003 and 2006) showed significantly higher NOx emissions rates than MOVES estimates of those vehicles.
- MOVES generally overestimates CO, NOx, and THC emission rates for Classes 6 and 4 vehicles.
- MOVES underestimates fuel consumption and CO_2 emissions of Class 4 HEs while MOVES overestimates those of Class 6 (normal emitting) vehicle.
- PM emission rates of all vehicles were significantly lower than MOVES estimates regardless of vehicle class/type probably due to different measurement methodologies between MOVES data (filter samples on dynamometer testing) and the test data in this study (light scattering detection during on-road PEMS testing).

Idling testing was conducted in TTI's EERF under a controlled test condition 86°F and 60 percent RH at one (low) engine speed, which was designed for comparisons among vehicle classes/types. During the idling tests, PM filters and MSAT cartridge samples and continuous mass monitoring were also taken in addition to PEMS emission measurements. The major findings of the idling testing are:

- Class 8 HEs showed the greatest fuel consumption and emission rates compared to the vehicles in other classes/types.
- For fuel consumption and CO_2, NOx, and CO emissions, Class 8 randomly selected vehicles were the next highest followed by Class 4 HEs and then Class 6 HDDVs.
- For PM filter and DMM measurement results, Class 6 HDDVs were the next highest followed by Class 4 HEs and then by Class 8 randomly selected HDDVs.
- For THC and aldehydes, Class 8 randomly selected vehicles were the next followed by Class 6 HDDVs and then Class 4 HEs.
- Compared to MOVES emissions rates, Class 8 HEs showed higher rates for CO, THC, and aldehydes and Class 4 HEs for CO_2 while all other measured emissions rates from Classes 8 and 4 HEs and all of the measured emissions from Class 8 randomly selected and Class 6 HDDVs were, on average, lower than the MOVES estimates.

Using the measured emission results and MOVES estimates, potential emissions benefits by replacing HEs were estimated as follows:

- Based on idling testing results, 21 percent of fuel savings and emission reductions of 18 percent (CO_2), 24 percent (NOx), 61 percent (CO), and about 80 percent (PM, THC, and aldehydes), on average, are anticipated by replacing Class 8 HEs with Class 8 randomly selected (MY 2003 & 2007) vehicles.
- Based on driving testing results and MOVES estimates, replacing an old vehicle (MY 2000 or earlier) with a vehicle that complies with 2011 emissions standards will significantly reduce all emissions in all three vehicle classes; about 8–12 percent of CO_2 reduction and at least 51 percent (up to 95 percent) of emission reductions of NOx, CO, THC, and PM.

The above results enabled the research team to characterize the emissions of vehicles identified as potential high emitters relative to non-high emitting vehicles. The test results were also compared to MOVES emissions rates. The findings regarding emissions of potential high emitter vehicles were applied in conjunction with MOVES to develop an algorithm that can quantify the emissions reductions associated with replacing or upgrading high emitter vehicles. Applying this algorithm can assist fleet managers and environmental agencies in estimating emissions benefits of upgrading an old vehicle or potential high emitter vehicle. The proposed methodology provides three levels of benefits based on the level of upgrade. The algorithm is provided in Appendix B along with an application example. A similar algorithm may also be constructed using other emissions models such as MOBILE6. However, since the MOBILE6 model is being phased out and replaced with MOVES, an algorithm compatible with MOBILE6 was not developed, consistent with the other analyses presented in this report, which also focus on MOVES.

Different PM measurement methodologies (opacity, filter, laser scattering, and ELPI measurement methods) were used in order to screen HEs and to measure PM emissions, and the measured PM results were compared. Among the PM measurement methods, some correlations were found. For example, with linear regressions, PM measurement results with using filter and ELPI methods showed some correlations for Class 8 HEs ($R^2 = 0.764$) and for Class 8 randomly selected vehicles ($R^2 = 0.858$). For comparisons with opacity readings, the researchers found almost no relationships between the opacity and filter method mainly due to different test conditions; opacity method was used during the high idling at HE selection in COH facilities while filter method was used during the low idling at idling testing inside EERF. More systematic studies with various test conditions would facilitate these findings.

The results from this project demonstrate that a viable emissions reduction strategy could be to screen high emitter vehicles from the fleet and replace them or install emissions control technologies for maximizing the emissions reduction and air quality benefits, through scrappage or retrofit programs. Larger vehicle fleets, especially those with older vehicles in the HGB area and other NA areas can provide many opportunities to apply these strategies for regional air quality improvement. While the study's findings suggest that a scrappage program for the replacement of high emitting vehicles would be beneficial to the air quality improvement in the HGB area or other NA areas, additional economic analyses can help determine if such programs are also cost effective and economically viable.

The following are the research team's recommendations for future investigation regarding screening and scrappage of HEs:
- There is a need to investigate and develop more robust methodologies for the identification/screening of high emitting vehicles.
- There is the potential to study more fleets to better characterize emissions and identify target fleets that maximize emissions benefits.
- The use of emission rates based on Texas-specific (or area-specific) drive cycles rather than rates based on national drive cycles would provide more accurate estimates for emission rates from a local perspective.

REFERENCES

1. Texas Commission on Environmental Quality (TCEQ). Houston-Galveston-Brazoria and the State Implementation Plan. http://www.tceq.texas.gov/airquality/sip/hgb/sip-hgb, accessed March 2008.

2. EPA, Final Rule on In-Use Testing Program for Heavy-Duty Diesel Engines and Vehicles. EPA420-F-05-021, June 2005.

3. Texas Commission on Environmental Quality (TCEQ). State Implementation Plan. http://www.tceq.state.tx.us/assets/public/implementation/air/sip/miscdocs/HGB_fact_sheet011906.pdf, accessed March 2008.

4. Texas Department of Public Safety. AirCheck Texas. Emissions Testing Procedures. http://www.txdps.state.tx.us/vi/inspection/item_insp.asp, accessed March 2008.

5. TCEQ. Development of 2008 CERR On-Road Mobile Source Actual Annual and Summer Season Weekday Emissions Inventories for the Houston-Galveston-Brazoria Area (Umbrella Contract 582-06-70880 / Work Order 90400-FY09-01: Task 2) - Final. Austin, TX. August 2009.

6. TCEQ. Is your car or truck 10 years old or older? Or has it failed an emissions test? http://www.tceq.texas.gov/airquality/mobilesource/vim/driveclean.html, accessed June 2011.

7. TCEQ, SIP Revision: Houston-Galveston-Brazoria 2007, Appendix A. http://www.tceq.texas.gov/assets/public/implementation/air/sip/hgb/hgb_sip_2007/appendices/06027SIP_Appendix_A_VMEP.pdf, accessed 2012.

8. TCEQ, TERP Emissions Reduction Incentive Grants (ERIG) Program. http://www.tceq.state.tx.us/implementation/air/terp/index.html, accessed March 2008

9. EPA. Mobile Source Air Toxics. http://www.epa.gov/OMSWWW/toxics.htm, accessed March 2008.

10. TCEQ. Houston Exposure to Air Toxics Study (HEATS). http://www.tceq.state.tx.us/implementation/tox/research/heats.html, accessed March 2008.

11. EPA. MOBILE6 Vehicle Emission Modeling Software. http://www.epa.gov/otaq/m6.htm, accessed March 2008.

12. EPA. Motor Vehicle Emission Simulator Highway Vehicle Implementation (MOVES-HVI) Demonstration Version – Draft Software Design and Reference Manual. EPA420-P-07-001, February 2007.

13. EPA. Draft Design and Implementation Plan for EPA's Multi-Scale Motor Vehicle and Equipment Emission System (MOVES). EPA420-P-02-006, October 2002.

14. Draft Software Design and Reference Manual: Motor Vehicle Emission Simulator Highway Vehicle Implementation (MOVES-HVI) Demonstration Version. MOVES Homepage. U.S. Environmental Protection Agency, February 2007. Available online at http://www.epa.gov/otaq/ngm.htm.

15. Lindhjem, C., A. Pollack., R. Slott., and R. Sawyer. Analysis of EPA's Draft Plan for Emissions Modeling in MOVES and MOVES GHG. *CRC Project E-68*. Report Prepared for the Coordinating Research Council, Inc. Environ, Novato, CA, May 2004.

16. Younglove, T., G. Scora, and M. Barth. "Designing On-Road Vehicle Test Programs for the Development of Effective Vehicle Emission Models." *Transportation Research Record.* 2005, Vol. 1941, pp. 51–59.

17. Zietsman, J., J. Villa, T. Forrest, and J. Storey. Estimating Truck Emissions at the El Paso – Ciudad Juarez Border. *Proceedings of the National Urban Freight Conference*, Long Beach, CA. February 2006.

18. Storey, J., S. Lewis, J. Zietsman, J. Villa, and T. Forrest. Mobile Source Air Toxics from Idling Trucks – A Report From The Mexican Border. *Preprint of the 86th Annual Meeting of the Transportation Research Board, National Research Council*, Washington, D.C. 2006.

19. California Air Resources Board. Proposed Statewide Diesel Truck and Bus Regulation. http://www.arb.ca.gov/msprog/onrdiesel/documents/Jan_Feb_2008_Workshops_Presentation.pdf, accessed March 2008.

20. EPA. Control of Emissions from New and In-Use Highway Vehicles and Engines. Code of Federal Regulations Title 40, Part 86. http://www.access.gpo.gov/nara/cfr/waisidx_04/40cfr86_04.html, accessed March 2008.

21. Caddle, S., P. Mulawa, E. Husanger, K. Nelson, R. Raggazi, R. Barret, G. Gallagher, D. Lawson, K. Knapp, and R. Snow. Light-Duty Motor Vehicle Exhaust Particulate Matter Measurement in the Denver, Colorado, Area. Journal of the Air & Waste Management Association. Vol. 49, pp. 164-174. September 1999.

22. SAE. Snap Acceleration Smoke Test Procedure for Heavy-Duty Powered Vehicles. SAE J1667, February 1996.

23. EPA. MOVES 2010 Users Guide. EPA-420-B-09-041, December 2009.

24. EPA. Development of Emissions Rates for Heavy-Duty Vehicles in the Motor Vehicle Emissions Simulator (Draft MOVES 2009). EPA-420-P-09-005, August 2009.

APPENDIX A: DRIVING EMISSIONS RESULT COMPARISONS WITH MOVES ESTIMATES

Table A-1. Percentage of MOVES Estimation (Class 8 Potential HEs).

Vehicle #	MY	Number of Cycle Sets	CO_2 (%)	CO (%)	NOx (%)	THC (%)	PM (%)
8H-1	1993	8	7.9%	−32.7%	−13.6%	467.9%	
8H-2	1994	8	16.0%	−9.1%	−17.2%	113.1%	−74.5%
8H-3	1995	8	14.3%	−28.7%	40.4%	10.8%	−77.4%
8H-4	1996	8	31.4%	−30.0%	17.6%	344.9%	
8H-5	2003	8	−12.6%	47.2%	44.5%	−31.8%	−86.5%
Average			11.4%	−10.7%	14.4%	181.0%	−79.5%

Table A-2. Percentage of MOVES Estimation (Class 8 Randomly Selected Vehicles).

Vehicle #	MY	Number of Cycle Sets	CO_2 (%)	CO (%)	NOx (%)	THC (%)	PM (%)
8-1	2003	8	−6.8%	−29.2%	39.1%	−7.7%	
8-2	2003	8	8.8%	14.2%	69.4%	−26.3%	−70.0%
8-3	2003	8	4.8%	65.1%	66.5%	−37.6%	−80.8%
8-4	2003	8	7.6%	34.5%	81.0%		−65.7%
8-5	2003	8	−2.7%	−3.5%	70.5%		−65.3%
8-6	2003	8	14.0%	4.9%	102.5%	−45.9%	−81.0%
8-7	2003	8	8.3%	−10.9%	80.3%	−63.8%	−80.8%
8-8	2003	8	9.6%	131.4%	69.2%	−66.5%	−62.7%
8-9	2003	8	12.8%	207.9%	97.6%	−72.2%	−65.6%
8-10	2006	8			33.1%		
8-11	2006	8	2.5%	33.4%	23.0%	−26.7%	−78.7%
8-12	2006	8	11.0%	57.1%	39.3%	−17.3%	
Average			6.3%	45.9%	64.3%	−40.4%	−72.3%

Table A-3. Percentage of MOVES Estimation (Class 6 Vehicles).

Vehicle #	MY	Number of Cycle Sets	CO_2 (%)	CO (%)	NOx (%)	THC (%)	PM (%)
6-1	1994	5	−83.1%	−89.4%	−87.5%		−92.5%
6-2	1999	5	−35.1%	−52.5%	−33.6%	−36.5%	−83.9%
6-3	2000	5	−24.2%	−26.7%	−27.6%	−26.9%	−80.9%
6-4	2000	5	−32.2%	−50.2%	−36.6%	−32.6%	−92.1%
6-5	2001	5	−71.9%	−76.1%	−78.3%	−92.4%	−91.1%
6-6	2002	5	−18.2%	−3.3%	4.1%	−77.9%	−86.3%
		Average	**−44.1%**	**−49.7%**	**−43.2%**	**−53.3%**	**−87.8%**

Table A-4. Percentage of MOVES Estimation (Class 4 Potential HEs).

Vehicle #	MY	Number of Cycle Sets	CO_2 (%)	CO (%)	NOx (%)	THC (%)	PM (%)
4H-1	1994	6	67.8%	−76.6%	−54.2%	−65.0%	
4H-2	1999	6	120.1%	−61.9%	−20.3%		−65.2%
4H-3	1999	6	32.8%	−76.8%	−39.1%		
4H-4	1999	6	100.9%	−62.0%	−28.3%	−97.1%	−55.2%
4H-5	1999	6	0.3%		−50.3%	−97.5%	−39.5%
4H-6	1999	6	6.5%	−82.4%	−49.0%		−99.1%
		Average	**64.4%**	**−69.3%**	**−38.5%**	**−86.6%**	**−53.3%**

APPENDIX B: MOVES-BASED ALGORITHM FOR ESTIMATING THE BENEFITS OF UPGRADING A POTENTIAL HIGH EMITTER VEHICLE

This proposed algorithm is developed based on a combination of observed emissions rates (obtained from the emissions testing results) and MOVES-generated emission rates. The baseline values developed in this algorithm are considered representative of the high emitter vehicles. These baseline values are constructed using average observed values supplemented by modeled emissions and are not linked to a specific model year. However, it is generally assumed that this algorithm would be applied for estimating potential emission reduction benefits from upgrading heavy-duty vehicles older than MY 2000, since these are most likely to be high emitters.

Assumptions:
- The algorithm covers three vehicle classes: class 8 (type 61), class 6 (type 52), and class 4 (type 32).
- Three levels of upgrades are as follows:
    o Low Impact: an upgrade that is expected to marginally improve the emissions characteristics of older vehicles. This impact level is simulated by assuming that the selected upgrade will bring the average emissions level of the vehicle to a level comparable with a MY 2003 vehicle.
    o Medium Impact: an upgrade that is expected to modestly improve the emissions characteristics of older vehicles. This impact level is simulated by assuming that the selected upgrade will bring the average emissions level of the vehicle to a level comparable with a MY 2006 vehicle.
    o High Impact: an upgrade that is expected to bring the emissions characteristics of an older vehicle to that of a brand new vehicle, i.e., MY 2011.
- The rates and adjustment factors provided in this algorithm are representative of general reduction in emissions levels for all pollutants, based by a study of overall differences between high-emitter and non-high emitter vehicles. However, in applying this algorithm to specific vehicle upgrades and retrofits, users must take account of individual factors that may affect the estimates of emissions reduced. For example, a selective catalytic reduction (SCR) has a high impact in terms of NOx reductions, but has no impact on other pollutants. In such cases, the algorithm will need to be applied in a selective manner for specific pollutants.
- Baseline emission rates (g/mi) are calculated based on the overall observed drive cycles of the City of Houston's vehicles that were utilized in the analyses of this study.

Methodology:
Step 1: Determine the following information for all the vehicles to be upgraded:
- Average annual mileage of vehicles (AAM);
- The type of the intended upgrade.
- The impact level of the intended upgrade based on its characteristics, i.e., low, medium, and high.

Step 2: For each vehicle use Equation B-1 to estimate the expected annual emissions reduction as the result of the intended upgrade, that is, total emissions reduced (TER).

Values for Baseline Emissions Rates (BER) and Emissions Adjustment Factor (EAF) are listed in Tables B-1 and B-2.

$$TER_i = BER_i \times AAM \times EAF_i ; \quad i = CO2, CO, NOx, THC, PM \quad (B-1)$$

Step 3: Repeat Step 2 for all vehicles being upgraded.
Step 4: Sum the individual estimated reductions for all vehicles.

Table B-1. Baseline Emissions Rates (BER_i).

	Baseline Emissions Rate (g/mi)				
	CO_2	CO	NOx	THC	PM
Class 8 (Type 61)	1735	5.259	17.252	3.990	0.184
Class 6 (Type 52)	936	2.755	6.011	0.822	0.086
Class 4 (Type 32)	1382	2.738	5.961	0.288	0.132

Table B-2. Emissions Rates Adjustment Factor (EAF_i).

	Level of Upgrade	CO_2	CO	NOx	THC	PM
Class 8 (Type 61)	Low	3%	6%	25%	80%	3%
	Medium	3%	9%	44%	80%	28%
	High	8%	92%	95%	97%	88%
Class 6 (Type 52)	Low	N/A	3%	10%	1%	15%
	Medium	1%	25%	34%	28%	27%
	High	1%	84%	86%	84%	80%
Class 4 (Type 32)	Low	14%	42%	15%	84%	24%
	Medium	14%	46%	35%	85%	24%
	High	20%	57%	86%	85%	92%

Example Application of Algorithm
A fleet owner wants to upgrade two of the fleet vehicles that are identified to be high emitters, as follows:
- Vehicle 1: a 1995 diesel tractor-trailer (class 8) with an average annual mileage of 50,000 miles over the past three years to be replaced with a brand new vehicle.
- Vehicle 2: a 1998 light delivery truck (class 4) with an average annual mileage of 12,000 miles to receive a comprehensive maintenance and repair service.

The expected upgrade for vehicle 1 is a high impact upgrade while the repair service for vehicle 2 is expected to provide low level emission benefits and is classified as a "low impact" upgrade. The expected CO_2 emissions reductions as results of these planned upgrades are calculated below:

Vehicle 1:
Annual CO_2 reduction = 1735 (g/mi) × 50,000 (mi/year) × 0.08 ÷ 1000 (convert to kg) = 6940 (kg/year)

Vehicle 2:
Annual CO_2 reduction = 1382 (g/mi) × 12,000 (mi/year) × 0.14 ÷ 1000 (convert to kg) = 2322 (kg/year)

Similar calculations can be performed for estimating reduction of other pollutant emissions by using the specific baseline rates and adjustment factors.

www.ingramcontent.com/pod-product-compliance
Lightning Source LLC
Chambersburg PA
CBHW081843170526
45167CB00007B/2887